CHANGING ROLES IN NATURAL FOREST MANAGEMENT

To Rupert, Lana, Tegan, Mam, Dad, Siân and Barry,
For life as it has been and life as it is now.

Changing Roles in Natural Forest Management

Stakeholders' roles in the Eastern Arc Mountains, Tanzania

KERRY A. WOODCOCK

LONDON AND NEW YORK

First published 2002 by Ashgate Publishing

Reissued 2018 by Routledge
2 Park Square, Milton Park, Abingdon, Oxon OX14 4RN
711 Third Avenue, New York, NY 10017, USA

Routledge is an imprint of the Taylor & Francis Group, an informa business

A Library of Congress record exists under LC control number: 2002019830

ISBN 13: 978-1-138-72850-9 (hbk)
ISBN 13: 978-1-138-72847-9 (pbk)
ISBN 13: 978-1-315-18997-0 (ebk)

Contents

List of Boxes

List of Figures

List of Tables

Preface

Development and environmental management issues in Africa are surrounded by ideological confusion. Ideological confusion surfaces in debates between optimists and pessimists about the surfeit or death of opportunity generated by African people in their environment. Hidden in the ideological confusion are ill focused debates about the impact of colonisation, meaning of independence and Africa's role in the globalisation process - cultures, history, politics and economics underlie claims to optimism or pessimism.

In the environmental debates issues sink and resurface, soil erosion, deforestation, drought, flood, food security and primary health care. The optimists and pessimists emerge again seeing local people as the cure or the curse respectively. As the last century drew to a close, two dominant global problems came to the fore, namely climate change and biodiversity that have significant implications for Africa's environmental future, a future made more uncertain by Africa's continuing poverty.

The maintenance of African biodiversity was again a battleground of ideological confusion. Crude conservation limited local development opportunity: crude development destroyed habitats and associated flora and fauna. The rallying call became "What works well?" essentially a call that sought to minimise the role of the state, seek new modalities of intervention through NGOs and pursue proven management techniques. All this "What works well?" agenda was based on minimising history, politics and economics.

In the last decade there has been an increasing focus on the management of Eastern Arc forests for biodiversity conservation. There has been general agreement that past management approaches have failed to manage the forests sustainably and increasing recognition that a greater role is required for forest-local communities. There is less agreement as to what that role should be. Research in the Eastern Arc has so far focused on the role of forest-local communities via their returns from forest resources. Little research has been conducted into forest-local communities' roles via their rights, responsibilities and relationships to forest.

Kerry Woodcock's work examines changing roles in natural forest management in the Eastern Arc Mountains, Tanzania. Stakeholders' roles are analysed via their respective rights, responsibilities, returns from forest resources and relationships, through changing natural forest management approaches. Imbalances between and within stakeholders' roles reflect unequal relations between stakeholders, which have contributed to unsustainable forest management practices. Balancing stakeholders' roles and empowering stakeholders to negotiate their respective roles will contribute to the development of sustainable forest management practices in the Eastern Arc Mountains.

This work, originally produced as her doctoral thesis, was based on Kerry Woodcock's fieldwork with two NGOs: Frontier-Tanzania and the Tanzania Forest Conservation Group. The methodology uses a multidisciplinary, case study approach combining three types of techniques: Participatory techniques, Ethnography and Secondary data analysis.

This volume amply demonstrates that good management requires an understanding of local culture, knowledge of history, sound economics and most of all empowerment. It is impossible to answer the question "What works well?" without building an inclusive politics.

Phil O'Keefe
Newcastle upon Tyne

Acknowledgements

I must first and foremost thank all those I have learned from and with in the process of this research, particularly those forest-local communities of Kambai, Kwezitu, Seluka, Mkwajuni, Kwatango, Mgambo, Mwembeni, Magula, Gare and Lulanda. There are several people from these communities that assisted me specifically in carrying out my research that I wish to thank in writing: Chrispin Sylvester Kamote, Benjamin Maua, Mwanamkuu Mohammedi and Valentin J. Kimwaga and Lukiano Tupa who have both sadly since passed on.

I thank those that I have worked with both in the Tanzania Forest Conservation Group (TFCG) and Frontier-Tanzania. Particular thanks go to Alex Hipkiss, Andrew Perkin and Aidano Makange of TFCG. I have learned a lot about negotiation from Makange and without his assistance life in a rural Tanzanian village could have been difficult at times. Makange deserves a doctorate for his communication abilities alone and would I believe, in another place and time, have achieved one. I am happy that his family will see this book, although Makange is no longer with us. I wish to thank my friend and neighbour in Kambai, Mary Christopher, who as my friend taught me how to live in and run a rural Tanzanian household and for being there when I needed a friend to talk to. Thanks go to my good friend Camilla Bildsten for the many hours of discussion surrounding this subject.

I, of course, thank Phil O'Keefe and John Kirkby for their supervision of the PhD upon which this book is based.

Last and not least I thank my family and friends for supporting me in all ways throughout this work, especially my parents and Rupert.

Asanteni sana!

List of Abbreviations

4Rs	Rights, Responsibilities, Returns, Relationships
CFM	Community Forest Management
CGFR	Central Government Forest Reserve
CM	Collaborative Management
CPR	Common Property Regime
CTA	Chief Technical Advisor
DC	District Commissioner
DFO	District Forest Officer
EUCADEP	East Usambara Conservation and Agricultural Development Programme
EUCFP	East Usambara Catchment Forest Project
FA	Forest Attendant
FAO	Food and Agriculture Organisation
FBD	Forestry and Beekeeping Division
FINNIDA	Finnish Aid Agency
FPS	Finnish Forest and Park Service
ICDP	Integrated Conservation and Development Programme
IUCN	World Conservation Union
JFM	Joint Forest Management
KFCP	Kambai Forest Conservation Project
LFCP	Lulanda Forest Conservation Project
LGFR	Local Government Forest Reserve
MNRT	Ministry of Natural Resources and Tourism
NGO	Non-Governmental Organisation
NTFP	Non-Timber Forest Product
PFM	Participatory Forest Management
PLA	Participatory Learning and Action
RRA	Rapid Rural Appraisal
SHUWIMU	Shirika la Uchumi la Wilaya ya Muheza (Muheza District Development Corporation)
SSI	Semi-Structured Interviewing
SSM	Sikh Sawmills
TANU	Tanzanian African National Union
TFCG	Tanzania Forest Conservation Group
TFP	Timber Forest Products
TWICO	Tanzania Wood Industries Corporation

UNEP	United Nations Environment Programme
UNICEF	United Nations International Children's Emergency Fund
US	United States
VDC	Village Development Committee

Please note that words in text that are:

- Bold-italic, are ***Kiswahili***;
- Bold, are **Kisambaa** or **Kihehe**; and
- Italic, are *Latin*.

Chapter 1

Biodiversity Conservation and Natural Resource Management

Evolution of Approaches to Forest Management in Africa

Natural resource management and biodiversity conservation in particular have become increasingly significant to both international and national communities over the last two decades. The paramount question is how to develop sustainable management approaches. For decades this question was addressed by many people who thought only in terms of two alternatives: markets or centralised government (Taylor 1997). More recently a third alternative has been offered by others (for example, Baland & Platteau 1996; Poffenberger & McGean 1996) who have argued the role of community in the management of natural resources.

Borrini-Feyerabend (1996), Chambers (1994b), Dubois (1997), Korten (1984), Pimbert and Pretty (1997) and Wily (1997 & 1999) have all analysed changes in approaches to natural resource management and biodiversity conservation to differing extents. Dubois (1997) and Wily (1997 & 1999) have specifically analysed the evolution of approaches to forest management in Africa, natural forest management being the specific focus of this research. Dubois (1997) argues concisely that in recent decade's forest management has passed through two main phases and is now entering a third:

- The technocratic era: management for the forest and against the people;
- The participatory era: forest management for and by the people; and
- The emergence of political negotiation: forest management with the people and other actors.

Although Dubois (1997) identifies an evolution of forest management approaches as three phases or eras of management, it is important to note that the evolution of forest management is perhaps not so rigid periodically and rather more dynamic than the term 'era' suggests. Although there are perhaps predominant management paradigms in particular periods, different management approaches may in fact co-exist. Table 1.1 summarises biodiversity conservation and natural resource management paradigms and contrasts the technocratic, participatory and political negotiation approaches.

Table 1.1 Biodiversity conservation and natural resource management paradigms: Contrasting technocratic, participatory and political negotiation approaches

	Technocratic	Participatory	Political negotiation
Approach	Blueprint	Process	Negotiation
Point of departure	Nature's diversity and its potential commercial values.	Nature's diversity and people's potential to protect or harm it.	The diversity of both people and nature's values.
Keyword	Strategic planning.	Participation.	Negotiation and collaboration.
Locus of decision making	Centralised, ideas originate in capital city or Northern countries.	Predefined problems and solutions originate in capital city or Northern cities. Decisions are made on predefined choices locally.	Decentralised, ideas originate from all stakeholders and decisions are negotiated.
First steps	Data collection and planning.	Awareness and action.	Stakeholders identified through open discussion and mechanisms for negotiation developed.
Design	Static, by experts.	Evolving, people involved.	Evolving: negotiated by stakeholders.
Main resources	Central funds and technicians.	Donor funds, local people and their assets.	Stakeholders' assets pooled.
Methods, rules	Standardised, universal fixed package.	Predefined basket of choices.	Diverse, local, varied basket of choices, evolving.
Analytical assumptions	Reductionist, natural sciences bias.	Pluralistic, exercise of judgement, economic bias.	Systemic, holistic, socio-political bias.

Management focus	Spending budgets, completing projects on time.	Product focus: e.g. interested in number of trees planted in one year.	Collaboration focus: sustained improvement and performance.
Communi-cation	Vertical: orders down, reports up.	Lateral: but on subjects defined from the 'top'.	Lateral: mutual learning experience.
Evaluation	External, intermittent.	External, donor led.	Internal, continuous.
Error	Buried	Embraced	Embraced
Relationship with local people	Controlling, policing, inducing, motivating, dependency creating, people seen as problem. Management for the forest and against the people.	Involving, people seen as beneficiaries. Forest management for and by the people.	Enabling, supporting, empowering, people seen as actors. Forest management with the people and other actors.
Interventions	Reservation	Farm forestry, multiple resource use, buffer zones, improved land use management.	Community forest management, joint forest management.
Outputs	1. Diversity in conservation, and uniformity in production (agriculture, forestry). 2. The empowerment of professionals.	1. Diversity as a principle of production and conservation. 2. The involvement of rural people.	1. Diversity as a principle of production and conservation. 2. The empowerment of rural people.

Source: Modified from Korten 1984, in Pimbert & Pretty 1997, with reference to Borrini-Feyerabend 1996, Chambers 1994b, Dubois 1997, Wily 1997 & 1999, and Author's fieldwork 1994-1998.

The Technocratic Approach

Two main strategic models for protected area management emerged in the 1960s and 1970s: exclusive management, which excluded local people; and inclusive management, which included local people (West & Brechin 1991: cited in Borrini-Feyerabend 1996). In the first - largely adopted in the US - management plans were developed with the intention of de-coupling the interests of local people from protected areas, with options ranging from an open anti-participatory attitude to the outright resettlement of the resident communities. This stance was common to both state-owned and privately owned reserves, including territories bought by conservation NGOs to prevent their exploitation by private developers. In the second model - more frequently adopted in Western Europe - the interests of local societies were central to the protected area: *the well-being of those who live and work in the National Parks must always be a first consideration* (Harmon 1991: cited in Borrini-Feyerabend 1996: 5). Private ownership of land within protected areas was common and local administrators were largely involved in management planning.

While an exclusive management approach is generally successful in preserving areas of wilderness and scenic beauty, the inclusive approach is obviously the model of choice for protected areas that include human residents and affect local livelihood in important ways. With or without the explicit intention of following the US experience, it is the former model that spread most extensively in the countries of the South, regardless of the social context in which the protected areas were being developed (Borrini-Feyerabend 1996).

The North imposed the exclusive model of management on Africa well before the 1960s and 1970s, with the colonial administrations being the locus of decision-making (Table 1.1). The point of departure was nature's potential commercial values, and it was the earlier missionaries and later the colonial Forestry Department Officials who were responsible for collecting the data on which plans for the gazettment of forest reserves were made (Table 1.1). Communication was 'top-down' and the relationship between the Forestry Department and local people was that of controlling, policing and inducing (Table 1.1). The local people were seen as the problem and priority was given to the trees at the expense of the people who used them (Dubois 1997). It was thought that enhanced technical capacity in forest management would be sufficient to guarantee their renewal for the good of the nation. Programmes aimed at developing capacity primarily concerned technical matters and were intended for government staff who were regarded as the experts (Table 1.1). Over the years, significant failures of 'top-down' initiatives, driven solely by technical considerations and from the top led to the realisation that failures in forest management were not due to lack of technical skills alone (Dubois 1997).

The Participatory Approach

Dubois (1997) argues that it is the flaws of the technocratic approach that have led to the pursuit of the concept and practice of participation. Participation is a means to ensure that local people's interests and needs are taken into account in the decisions concerning the fate of forests (Table 1.1).

In the 1980s and 1990s parallel shifts of paradigm, supporting participation, have also been noted in four major domains of human experience: in the natural sciences, in the social sciences, in business management, and in development thinking (Chambers 1994b). All have in common, decentralisation, open communication and sharing knowledge, local diversity, personal responsibility and judgement, and rapid change. The word paradigm is used here to mean a coherent and mutually supporting pattern of concepts, values, methods and action, amenable to wide application.

In the natural sciences, conventional approaches, using hard systems and reductionist assumptions and methods, are in crisis when faced with many of our important problems (Mearns 1991; Appleyard 1992). Scientific method is not competent to predict or prescribe for the complex open systems that matter most. Global environmental issues involve huge uncertainties and demand what Funtowicz and Ravetz (1990: cited in Chambers 1994b) call a 'second order science' in which judgement plays a more recognised part. The science of ecology has questioned the validity of baselines such as a 'climax vegetation community', or static concepts such as 'carrying capacity' - notions which have been integral to the scientific validation of orthodox views of environmental change (Leach & Mearns 1996). Several studies have demonstrated the importance of using historical and 'time-series' data sets of various types, combining photographic with official written records, early travellers' accounts, and ethnographic research methods such as oral history, to study the process of landscape change more directly, thereby documenting history rather than inferring it (Fairhead & Leach 1996a & 1996b; Conte 1996; Kimambo 1996). Older theories are now yielding to greater pluralism in ecological thinking and method, through flexible and continuous learning and adaptation, and through the exercise of judgement (Table 1.1).

Attention to historical detail, and the shedding of theoretical straitjackets in the natural sciences, have converged with a better understanding of the land-use practices of Africa's farmers and herders, and of their own ecological knowledge and views of environmental change. Social anthropologists (Richards 1985; Fairhead 1992; Chambers et al 1989) provide this in the social sciences by recent work. The application of recent approaches in ecological history reveals the logic and rationality of indigenous knowledge and organisation in natural resource management. By contrast, received wisdom would have much of the blame for vegetation change perceived by outsiders as environmental 'degradation' resting with local land-use practices, whether labelling them as ignorant and indiscriminate or - more commonly - as ill-adapted to contemporary socio-economic and demographic pressures. In such accounts, rural people's ecological knowledge is notable mostly by its absence.

In business management, the parallel shift has been from the values and strategies of mass production to those of flexible specialisation. Standardisation has been replaced by variety and rapid response; hierarchical supervision by trust; and punitive quality control by personal quality assurance at source (Chambers 1994b). Tom Peter's book of advice to US business managers, *Thriving on Chaos: Handbook for a Management Revolution* (1987: cited in Chambers 1994b), advocates achieving flexibility by empowering people, learning to love change, becoming obsessed with listening, and deferring to the front line. The theme of local knowledge and action is also strong in *The Fifth Discipline: The Art and Practice of the Learning Organisation* (Senge 1992: cited in Chambers 1994b). Senge (1992: 228) writes:

> Localness is especially vital in times of rapid change. Local actors often have more current information on customer preferences, competitor actions, and market trends; they are in a better position to manage the continuous adaptation, which change demands.

In development thinking, normative theories of universal economic growth as the main means to a better life are no longer tenable (Ekins 1992). Economic growth ceases to be a simple, universal objective; it is recognised as environmentally harmful among richer countries; and economic resources are recognised as finite. There is a search for alternative normative paradigms, for more sustainable ways to enhance the quality of life. For the rich, the question is how to be better off with less; for the poor, it is how to gain more and be better off without repeating the errors of the rich. One way to serve these objectives is to enable local people to identify, express and achieve their own priorities. In line with this, the emergent paradigm for living on and with the Earth brings together decentralisation, democracy and diversity. What is local and what is different is valued. The trends toward centralisation, authoritarianism, and homogenisation are opposed. Reductionism, linear thinking, and standard solutions give way to an inclusive holism, open systems thinking, and diverse options and actions (Chambers 1994b).

A variety of institutional and legal frameworks exist in the modern developing world through which the involvement of local communities in forest management is expressed (Fisher 1995; Borrini-Feyerabend 1996; Wily 1997). Collaborative Management (CM) (also referred to as co-management, participatory management, joint management, shared-management, multi-stakeholder management or round table agreement) is the most common framework and is used to describe a situation in which some or all of the relevant stakeholders in a protected area are involved in a substantial way in management activities (Borrini-Feyerabend 1996). Some countries, most notably India and Nepal, have gone so far as to promulgate supporting legislation, providing an administratively and jurally bound framework for collaboration between state and people (Asia Sustainable Forest Network 1994; Talbott & Khadka 1994; Asia Forest Network 1995).

Borrini-Feyerabend (1996) argues that collaborative management is a continuum between 'actively consulting' and 'transferring authority and responsibility' (Figure 1.1). Figure 1.1 refers to possible de facto situations seen from the perspective of an agency in charge of a natural resource, regardless of underlying tenure rights, policies and legislation.

Others maintain that it is not appropriate to use the term 'collaborative management' for a situation in which stakeholders are merely consulted and not given a share of authority in management, and have proposed different terms for different levels of involvement (Franks 1995). Pretty (1994) has devised a typology of participation, which accounts for the different levels of participation (Table 1.2). Borrini-Feyerabend (1996: 16) argues however, that:

> It is difficult to identify a sharp demarcation between various levels of participation in management activities. For instance, a process of active consultation with local stakeholders may result in the full incorporation of their concerns into a natural resource management plan. Conversely, a lengthy negotiation in which various stakeholders hold seats in a decision-making body may leave many local demands unmet. Is the second necessarily more 'collaborative' management than the first?

In comparison to the pivotal experience of South Asia, community involvement in natural forest management in Africa has been slow to evolve (Shepherd 1991; World Bank 1992; Sharma et al 1994). Even now it is largely confined to discrete, and usually donor-prompted (Table 1.1), initiatives rather than arising from shifts in national policy or practice (Wily 1997).

However, as Wily (1997) points out, this does not mean that African

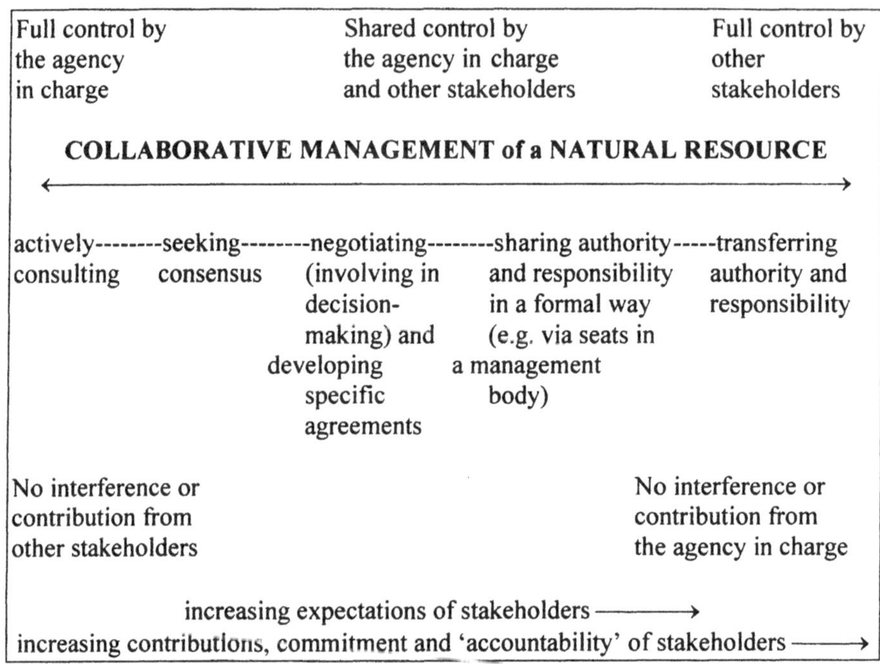

Source: After Borrini-Feyerabend 1996.

Figure 1.1 Participation in natural resource management - a continuum

governments have not been aware of the link between forest resources and people. Modern forest policies in many sub-Saharan states retain the commonly articulated principle of the British colonial forest policies that "*...the satisfaction of the needs of the people must always take precedence to the collection of revenue...*" (Government of Tanganyika 1945: cited in Wily 1997: 3). Such positions, locating people as forest users (Table 1.1), probably explain the strong orientation of most modern African community forestry initiatives towards reducing rather than increasing local vested interest in forests (Wily 1997).

In such initiatives, the forest-local community is either the target of investment to substitute local use of forest products (on-farm tree planting being the paramount intervention), and/or to displace forest-based economic dependence through promotion of improved agricultural production and off-farm income opportunities (Table 1.1). Buffer zone programmes and so-called Integrated Conservation and Development Programmes (ICDPs) were frequently developed on this basis in the 1980s (Table 1.1).

Table 1.2 A typology of participation

Typology	Components of each type
Passive participation	People participate by being told what is going to happen or what has already happened. It is unilaterally announced by an administration or by project management; people's responses are not taken into account. The information being shared belongs to only external professionals.
Participation in information-giving	People participate by answering questions posed by extractive researchers and project managers using questionnaire surveys or similar approaches. People do not have the opportunity to influence proceedings, as the findings of the research or project design are neither shared nor checked for accuracy.
Participation by consultation	People participate by being consulted, and external agents listen to views. These external agents define both problems and solutions, and may modify these in the light of people's responses. Such a consultative process does not concede any share in decision-making and professionals are under no obligation to take on board people's views.
Participation for material incentives	People participate by providing resources, for example labour, in return for food, cash or other material incentives. Much in-situ research and bioprospecting falls in this category, as rural people provide the resources but are not involved in the experimentation or the process of learning. It is very common to see this called participation, yet people have no stake in prolonging activities when the incentives end.
Functional participation	People participate by forming groups to meet predetermined objectives related to the project, which can involve the development or promotion of externally initiated social organisation. Such involvement does not tend to be at early stages of project cycles or planning, but rather after major decisions have been made. These institutions tend to be dependent on external initiators and facilitators, but may become self-dependent.
Interactive participation	People participate in joint analysis, which leads to action plans and the formation of new local groups or the strengthening of existing ones. It tends to involve interdisciplinary methodologies that seek multiple perspectives and make use of systematic and structured learning processes. These groups take control over local decisions, and so people have a stake in maintaining structures or practices.
Self-mobilisation	People participate by taking initiatives independent of external institutions to change systems. Such self-initiated mobilisation and collective action may or may not challenge existing inequitable distributions of wealth or power.

Source: Modified from Pretty 1994: in Pimbert and Pretty 1997.

Alternatively, the forest-local community becomes the target of benefit-sharing arrangements, in which a proportion of revenue is distributed from state to community, usually in the form of infrastructure from the wildlife sector, which has available a usually more financially-rewarding product namely revenue from hunting and tourism (Nhira & Matose 1996).

The attempt may also be made to confine the area of local forest use by designing multiple use zones (Table 1.1) on the periphery of the forest, where local people are permitted to harvest defined products, usually under rigorous supervision or conditions. In still other instances, efforts are made to 'meet the needs' of forest-local people through giving them priority in the issue of commercial licences to extract timber, poles or fuelwood.

These diverse strategies nonetheless share a product rather than a management focus (Table 1.1), and an understanding of the troubled state-people-resource relationship that is defined as simply a matter of economics (Wily 1997). This confirms the traditional assumption that such interests as forest-local people may have in a forest are solely product-based and that such conflicts as exist between state and people may be resolved solely by increasing access to the products. Issues of socio-cultural and especially customary tenurial rights tend to be ignored (Hobley & Shah 1996; Poffenberger et al 1996; Wily 1997). Indeed, economic determinism (Table 1.1) in conservation-driven natural forest management has reached unprecedented heights of elaboration over recent years. State of the art calculations of local forest values generate in turn inevitably expensive and unsustainable investment proposals for alternative income generation without resolving the causes of forest related free loader behaviour (Wily 1997).

Fundamentally, such approaches share a definition of the forest-local community as beneficiary (Table 1.1), not actor in resource management. What is being offered and shared in the common collaborative forest management approach is the use of certain forest products, not rights over the resource itself, nor the authority behind forest management (Wily 1997; Dubois 1997). This explains the research focus in the Eastern Arc Mountains, where this research is based, on returns from forest resources in the early to mid 1990s (Evers 1994; Kessey & O'Kting'ati 1994; Norton-Griffiths & Southey 1995; Owen 1992; Woodcock 1995).

Nor does involving local people in practical management tasks - usually surveillance - necessarily constitute a meaningful form of participation (see Table 1.2: functional participation). In such arrangements the local partner is often fulfilling protection activities according to the programme of the authority, such as the Forestry Department. Labour, or the burdens of management, rather than the rights of management, are being traded.

Programmes where an increase of responsibility is given to local people, without a corresponding increase in their rights, actually becomes a burden and is usually refused or passively accepted (Dubois 1997; Hobley & Shah 1996).

Although improved co-operation may be achieved, such approaches continue to avoid what Wily (1997) and Dubois (1997), regard as the more fundamental question to be addressed: the socio-political relations which drive state-people conflict and forest degradation. Viewed from this perspective, the foremost task becomes not the redefinition of who uses the forest and how, but of who owns, controls and manages the forest.

It is important to note, that although the predominant management paradigm at any one time may be viewed as participatory, often both technocratic and participatory approaches may be utilised simultaneously (for example, state forest reservation combined with implementations such as farm forestry).

The Emergence of Political Negotiation

What Dubois (1997) and Wily (1997) believe has been missing in both the technocratic and participatory approaches, is the recognition of the highly political character of forest management, even at local level (Table 1.1). Dubois (1997) highlights the need for a social definition of forest management, but stresses that this requires negotiations between institutions, which represent all existing interest groups and especially the weaker ones (Table 1.1). Hence, the implementation of collaborative forest management needs to be politically negotiated and mechanisms developed, which allow for the negotiation of stakeholders' roles (Dubois 1997; Hobley & Shah 1996).

In South Asia, particularly in India and Nepal, there have been a large number of examples of such approaches to forest management over the last 20 years, which have eventually been backed up by state policy and legislation (Talbott & Khadka 1994). Pattnaik & Dutta (1997) give a list of a number of papers concerning these approaches. Pattnaik & Dutta (1997: 3225) define Joint Forest Management (JFM) to be:

> where the owner (government), as well as the user (communities) jointly manages the forests and share the cost as well as the benefits. Whereas in community-based forest management (CFM), the community as the initiator is the sole protector and beneficiary of the forest with only moral support and a passive role of the government. And in participatory forest management (PFM) government is responsible for the forest protection with the paid help of the community where the benefits are shared, but the larger chunk of which is appropriated by the government.

Critiques of JFM have arisen in recent years. Pattnaik & Dutta (1997) state that JFM assumes an equal relationship between the government and local communities, yet more often than not it is the state, which retains the power. Power relations are again the main issue in Locke's (1999) critique of how gender is addressed in JFM policy in India. *Within JFM, social scientists concerned with gender need to advance theoretically-informed understandings of gender and the environment to go beyond telling the untold story of women living in the face of ecological distress and privileging women as a category above other participants in JFM* (Locke 1999: 282). What is required is an understanding of women's relationships of power and authority, negotiation and bargaining and the wider social relations in which decisions about the environment are made (Leach 1991: cited in Locke 1999).

Wily (1997: 4) argues that *an ideal transformation in power relations devolves authority to those within whose socio-spatial sphere the forest falls* (Table 2.1), *and who alone have the practical capacity to protect and supervise forest utilisation on a continuing basis.* The emphasis of forest use rights and responsibility shifts then to that of responsibility and authority. The basis of forest management is then what Wily (1997: 4) calls *a state-people collaboration in which the state supports the effort of the people rather than the people supporting the effort of the state.*

In comparison to South Asia, African states have been slow to evolve collaborative management approaches. In Tanzania specifically, examples have developed in the 1990s in the Miombo woodlands of Mgori and Duru-Haitemba (Wily 1996 & 1997). These are examples, which Wily (1999) believes move from JFM, where the state holds overall authority, into genuinely community-based approaches where local authority and responsibility of the forests is secured. It is not only the transfer of power from the state to community, but from community leaders to ordinary community members that these examples have demonstrated (Wily 1999). These examples of community-based management come from the Miombo woodlands, rather than high biodiversity forests, such as the Eastern Arc forests, where the state has not as yet entrusted local people with the guardianship of these forests. This is an issue which Wily (1999) believes demonstrates that community involvement in forest management is still viewed by the majority as a side issue to participatory management, designed to reduce conflicts between users rather than as an effective approach to forest management in its own right.

In March 1998 the new Tanzanian National Forest Policy was announced. Table 1.3 summarises key policy statements that refer specifically to stakeholders' involvement in the management of Tanzanian

Table 1.3 Key National Forest Policy statements

Forest status	Policy statement
Central and Local Government Reserves	Policy statement (3): To enable participation of all stakeholders in forest management and conservation, joint forest management agreements, with appropriate user rights and benefits, will be established. The agreement will be between the central government, specialised executive agencies, private sector or local governments, as appropriate in each case, and organised local communities or other organisations of people living adjacent to the forest.
Forests on public lands (non-reserved forest land)	Policy statement (5): To enable sustainable management of forests on public lands, clear ownership for all forests and trees on those lands will be defined. The allocation of forests and their management responsibility to villages, private individuals or to the government will be promoted. Central, local and village governments may demarcate and establish new forest reserves.
Private and community forestry	Policy statement (6): Village forest reserves will be managed by the village governments or other entities designated by village government for this purpose. They will be managed for production and/or protection based on sustainable management objectives defined for each forest reserve. The management will be based on forest management plans. Policy statement (7): Private and community forestry activities will be supported through harmonised extension service and financial incentives. The extension package and incentives will be designed in a gender sensitive manner.

Source: After the United Republic of Tanzania 1998.

forests. It is important to note that the policy was formalised just after the completion of the field research, although the policy had been in the process of being developed over ten years or more, so many drafts had been seen. The new policy moves towards more meaningful roles for forest-local communities, although the extent of their roles depends up on the official status of the forest in question. What is offered in both Central and Local Government Forest Reserves are joint forest management agreements (Table 1.3), whereas in the forests on public and private lands community forest management is a possibility (Table 1.3).

Joint Forest Management (JFM) in the Tanzanian National Forest Policy
The forest policy proposes that joint forest management agreements are made for Central and Local Government Forest Reserves. JFM is defined in the forest policy as the *involvement of local communities or NGOs in the management and conservation of forests and forest land with appropriate user rights as incentives* (The United Republic of Tanzania 1998: x). From this statement alone, the role of forest-local communities has been defined as predominantly that of beneficiaries of forest resources, where *user rights and benefits* (Table 1.3) have been emphasised. With the forest policy in place, there are greater opportunities for the involvement of forest-local communities in the management of the forest reserves. The emphasis on the role of forest-local communities is however, once again, as with the participatory approach, placed on access and user rights, the responsibility of the work of management and returns from forest resources rather than on management rights and responsibilities.

The negotiation process itself must however be brought into question. How can stakeholders' with unequal power successfully negotiate their roles in the management of the forest reserves? The danger is that the State may try to facilitate the negotiation process itself, a situation that could impair the success of the process and increase conflict between two groups of stakeholders with a long history of mistrust.

This is where perhaps the State should bring in independent facilitators of negotiation, who are trained and experienced in negotiation. Perhaps this could be a role for NGOs. The danger here is that some NGO staff may not have the ability or confidence to facilitate the negotiation process or worse still may not yet be fully convinced of the ability of forest-local communities to play a meaningful - if any - management role. This may not only be the case with some NGO staff, but also with State forestry staff even at senior levels, particularly those who originate from purely biological and technocratic forestry backgrounds. This is highlighted by the policy itself, which entrusts forest-local communities to manage forests on public and private lands, but does not fully entrust them with the management of the forest reserves, many of which are highly valued by the State and the International Community for their high biodiversity. The danger is that adequate resources and time will not be put into the process. If the negotiation process is not facilitated by independent experienced professionals who are aware of unequal power relations between stakeholders, then the capacity for negotiation of JFM agreements is likely to be minimal.

Community Forest Management (CFM) in the Tanzanian National Forest Policy The forest policy enables forest-local communities to own and manage, both forests on public and private land (Table 1.3), which is identified as community forest management. However, central and local government are also enabled to own and manage forests on public land through further reservation of forests (Table 1.3), a situation that could be seen as returning to technocratic approaches of the past.

In CFM, as in JFM, an understanding of the roles of and relations between stakeholders is required, although at community level. The forest-local community is not a homogenous unit and in any development of CFM, roles will need to be negotiated. Without an understanding of the institutional context of negotiations at local level, particularly imbalances in relations between village leaders and ordinary villagers, the village and sub-villages, men and women, young and old, immigrants and non-immigrants, then the capacity for negotiation could be minimal.

The practicalities of managing the forest also raise questions. Legally it is the village as a whole, through the village government who may be the owners of forests. However, as this research demonstrates (Chapter Four) customary ownership and management of forests was often held by clans within a larger settlement or village, now often in the form of sub-villages. Hence, issues of who can best manage the forest locally might be raised. Who is the forest-*local* community? Is it the village as legal owner of the forest or the sub-village, which is physically closest and may have perhaps held customary management authority in the past? Would authority and responsibility be hierarchical in terms of the village government or would there be a shift to the village assembly and ordinary villagers? Would men be more involved than women would? Legally what mechanisms would need to be put in place for ordinary villagers to fine or charge those not adhering to forest rules?

Independent facilitators of negotiation would again be required to assist community stakeholders to negotiate their roles. The NGOs could again take the role of facilitator and advisor, being wary of the project based approach, where communities may become dependent on the assistance of NGOs and the NGOs themselves may mistakenly perceive themselves as primary stakeholders. A facilitative, consultative role would seem to be more empowering to forest-local communities and more effective in terms of resources and the sustainability of forest management.

The State's role in community forest management is proposed by policy statement seven (Table 1.3) as offering a *harmonised extension service and financial incentives*. What the financial incentives would be and why they are seen to be required is a bit worrying. As far as the extension service is concerned this is perhaps where the State can be most useful in supporting the efforts of the community, through technical support in tree growing and boundary marking. It would also seem appropriate for State forestry staff and NGOs to work together in the facilitation process, once forest-local communities are certain the State does not wish to appropriate the forests for themselves. In this way, both the State and forest-local communities can begin to forge complementary partnerships and perhaps decades of mistrust can begin to fade.

The same dangers apply in the facilitation process as in Joint Forest Management, if facilitators are inexperienced or worse still do not believe in the ability of forest-local communities to manage forests. Perhaps one of the biggest constraints is that State forestry departments and NGOs alike often have mandates to manage only forests of high biodiversity. Since many of the forests on public and private land have tended to be the most seriously degraded, then often they are not particularly valued for their high biodiversity value. This could seriously jeopardise the success of community forest management if resources in the form of time and expert facilitators are not put into these forests. Without sustainably managed community forests there could be a knock-on effect as to the success of jointly managed forests.

Stakeholders in Natural Forest Management

The management or mismanagement of a natural resource affects various groups in society (Borrini-Feyerabend 1996). First among these groups are the communities who live within or close to an area of natural resource and, in particular, the people who use or derive an income from their natural resources. These people possess knowledge, capacities and aspirations that are relevant for management, and recognise in the area a unique cultural, religious or recreational value. Many such communities possess customary rights over natural resources, although official recognition of those rights may be uncertain or nil (Borrini-Feyerabend 1996).

In addition to local residents and resource users, other social actors may have an interest in natural resource management (Table 1.4). In particular, these actors include the governmental agencies dealing with a range of resources (for example, forests, freshwater, fisheries, hunting tourism, agriculture) and the administrative authorities (for example, district or regional councils) dealing with natural resources as part of their broader mandate. They include the local businesses and industries (for example, tourist operators, water users) that can be significantly affected by the status of the area of the natural resource. They include those research institutions and non-governmental organisations (for example, local, national or international groups devoted to environment and/or development objectives) which find the relevant territories and resources at the heart of their professional concerns.

The various institutions, social groups and individuals that possess a direct, significant and specific stake in an area of forest are referred to as its 'stakeholders'. The stake holding may originate from institutional mandate, geographic proximity, historical association, and dependence for livelihood, economic interest and a variety of other capacities and concerns. In general (Borrini-Feyerabend 1996):

- Stakeholders are usually aware of their interests in the management of the protected area (although they may not be aware of all management issues and problems);
- Stakeholders usually possess specific capacities (such as, knowledge skills) and/or comparative advantage (such as, proximity, mandate) for such management; and
- Stakeholders are usually willing to invest specific resources (such as, time, money, and political authority) in such management.

Not all stakeholders are equally interested in conserving a resource. Borrini-Feyerabend (1996) argues therefore that they are not equally entitled to have a role in resource management. For this reason, she feels it necessary to distinguish among them on the basis of some agreed criteria (Box 1.1). Social actors who score high on several accounts may be considered 'primary' stakeholders. 'Secondary' stakeholders may score high only on one or two. Therefore, in collaborative management processes, she argues that primary stakeholders would assume an active role, possibly involving decision-making (for example, holding a seat on a management board). Secondary stakeholders would be involved in a less important way (for example, holding a seat in a consultative body).

Stakeholders organised in groups and associations (for example, a village council) generally possess an effective representation system (Borrini-Feyerabend 1996). In other cases, the stakeholders cannot count on an institutional structure capable of conveying their interests and capacities in an effective manner. In fact, it is an unfortunate development of recent history that many communities who did posses traditional institutions for resource management have seen them devalued and weakened by modern state policies that do not recognise them nor assign to them any meaningful role (Baland & Platteau 1996; Bromley & Cernea 1989). In some cases, effective traditional systems of resource management still exist, but their

Table 1.4 Social actors potentially stakeholders in natural resource management

Actor	For example
Individuals.	Owners of relevant land holdings in the natural resource area.
Families and households.	Long-term local residents.
Traditional groups.	Extended families and clans, with cultural roots in the natural resource area.
Community-based groups.	Self-interest organisations of resource users, neighbourhood associations, gender or age-based associations.
Local traditional authorities.	A village council of elders, a traditional chief.
Local political authorities prescribed by national laws.	Elected representatives at village or district levels.
Non-governmental bodies that link different relevant communities.	A council of village representatives, a district level association of fishermen societies.
Local governance structures.	Administration, police, judicial system.
Agencies with legal jurisdiction over the natural resource area at stake.	A State Park Agency with or without local offices or an NGO set in charge by the government.
Local governmental agencies and services.	Education, health, forestry and agriculture extension.
Relevant non-governmental organisations at local, national and international levels.	Environment or development dedicated.
Political party structures at various levels.	
Religious bodies at various levels.	
National interest organisations/people's associations.	Workers' unions.
National service organisations.	The Lions club.
Cultural and voluntary associations of various kinds.	A club for the study of unique national landscapes, an association of tourists.
Business and commercial enterprises.	Local, national and international, from local co-operatives to international corporations.
Universities and research organisations.	
Local banks and credit institutions.	

Government authorities at district and regional level. National governments.	
Supra-national organisations with binding powers on countries.	The European Union.
Foreign aid agencies.	
Staff and consultants of relevant projects and programmes.	
International organisations.	UNICEF, FAO, UNEP.
International unions.	IUCN.

Source: After Borrini-Feyerabend 1996.

Box 1.1 Possible criteria to distinguish among stakeholders

- Existing rights to land or natural resources;
- Continuity of relationship (for example, residents versus visitors and tourists);
- Unique knowledge and skills for the management of the resources at stake;
- Losses and damage incurred in the management process;
- Historical and cultural relations with the resources at stake;
- Degree of economic and social reliance on such resources;
- Degree of effort and interest in management;
- Equity in the access to the resources and the distribution of benefits from their use;
- Compatibility of the interests and activities of the stakeholder with national conservation and development policies; and
- Present or potential impact of the activities of the stakeholder on the resource base.

Source: After Borrini-Feyerabend 1996.

communication with outsiders (and thus their recognition) is quite problematic (Borrini-Feyerabend 1996).

Two actors that are potential stakeholders in natural resource management are the local community and local NGOs, and are defined more fully in the following.

Local Community

In most cases, the basic stakeholders in forest are the people living within or adjacent to forest, usually grouped under the term 'local community' (or communities). Often these people are directly dependent on the forest resources for their livelihood, cultural identity and well being. Yet, communities are complex entities, within which differences of ethnic origin, class caste, age, gender, religion, profession and economic and social status can create profound differences in interests, capacities and willingness to invest for the management of the forest.

Local NGOs

Non-governmental organisations (NGOs) are often divided into those from the North, Northern NGOs and those from the South, Southern NGOs, since the political economy of each differs (see Cherrett et al 1995). Since this research is based in Africa, when referring to Southern NGOs they are termed 'local NGOs'.

Since the mid-1980s there has been a dramatic increase in the number and importance of local environment and development NGOs in Latin America, Asia and Africa (Yadama 1997). NGOs throughout the developing world have been increasingly entrusted with the responsibility to deliver development programmes as the state's responsibility is diminishing. The general rise in NGO involvement parallels a rise in criticism of state-sponsored development programmes and NGOs have, until recently, been accepted as being better able to implement development programmes (Yadama 1997). Debates have, however, arisen as to the real ability of NGOs to provide sustainable development (Cherrett et al 1995; Yadama 1997).

There is considerable debate over a suitable definition for NGOs, but Cherrett et al (1995) have divided them into two broad categories: grassroots and professional organisations. The former, are usually community-based and concentrate on the process by which their aims might be achieved. Professional development NGOs tend to be creations of intellectuals or professionals and are usually constructed to do a particular job. Cherrett et al (1995) have attempted a taxonomy of 'environmental' NGOs in Africa, which is summarised in Table 1.5.

They have noted changes in the roles of conservationist and environmentalist NGOs. Conservationist organisations have redefined themselves as environmental, while the environmental NGOs have redefined themselves as pro-sustainable development. The conservationists have become aware that the broad African public sees conservation as the concern of wealthy whites for the preservation of animals and the creation of Nature Reserves. The redefinition of conservation organisations is therefore seen by Cherrett et al (1995) as a function of self-preservation, in response to changes in both the national and international contexts.

The environmentalists have reassessed their previous emphasis on nature rather than people. In general, because they have been working with

Table 1.5 Taxonomy of African 'environmental' NGOs

Taxonomy	Characteristics
Govern-mental	These reflect current trends to transfer policy formulation from government to the private sector. They are a product of multilateral and bilateral funding, and usually set up from within government in response to donor indications. They are technology driven.
Entre-preneurial	Development aid has attracted its fair share of entrepreneurs, who see it as a 'soft' sector with weak management and gullible donors.
Networking	Pan-African or global networks are directly funded by multi lateral or bilateral agencies. Even where they have not been driven and designed by a donor, they have tended to originate in response to the needs of Northern NGOs, and respond to a Northern agenda.
Conserv-ationist	These have all been initiated with private-sector funding and their operational model is that of business. They look to close collaboration with governmental, multilateral, and major bilateral agencies. For those with a more extended membership base, it is predominantly white and middle class.
Advocacy	The primary focus of these organisations is on influencing policy. Advocacy groups tend to be driven by the need to achieve legitimacy among current policy makers, and their reference group tends to come from academia and government.
Environ-mental	These are involved in the implementation of projects and programmes in the field. As a result, they have become increasingly sensitive to the need for mechanisms to link them with the communities where they work, and to the constraints imposed by the lack of organisation among these same communities. Increasingly, they are emphasising the need for grassroots training, and redefining their focus on sustainable development rather than the 'environment', and on sustainable land use rather than sustainable agriculture. Their major source of funding is Northern NGOs, although the most professional of them are already implementing bilaterally funded projects and are increasingly being sought by major donors to execute their policies.
Community-based	These can be divided into two groups. The first are those which have emerged from a more traditional 'welfare' background. Their social base tends to retain welfarist attitudes of deference to authority and a level of passivity concerning the instruments of power. Their funding comes from Northern NGOs or occasionally national government. The second group, are more explicitly transformative, community-based organisations. These reflect a practical coalition between the grassroots and professionals. To date, their funding sources have been Northern NGOs.

Source: After Cherrett et al 1995 (reproduced with permission of Oxfam Publishing).

people, the crisis of poverty has been inescapable and so led to growing awareness of a need to redefine the starting point as people, rather than 'nature' (Cherrett et al 1995).

At the grassroots, community-based level the vision is more holistic as environment and society are directly linked. The task of the grassroots organisation is to build up the capacity of the people themselves to become actors in determining their own agendas. They are the key to people-centred development, where both national governments and international community serve as instruments of the people, and are accountable to them (Cherrett et al 1995). The role of the professional NGO is then in facilitating grassroots, community-based NGOs and institutions for sustainable development.

Stakeholders' Roles in Natural Forest Management

In recent decades, approaches to natural forest management in Africa have evolved. The stakeholders in forest resources have also changed through different management eras, along with their roles in forest management. Dubois (1997) shows concern for the vagueness associated to the term 'roles'. He suggests that this weakness can be overcome by defining stakeholders' roles via their respective rights, responsibilities, returns from forest resources and relationships ('4Rs'). Stakeholders' '4Rs' are often unbalanced a situation that often impairs adequate negotiation and leads to deforestation and forest degradation (Aluma et al 1996; Dubois 1997; Wily 1996). The next section takes each of these '4Rs' in turn and reviews the literature.

Relationships to Forest[1]

The relationship between two things (for instance, forest and people) or two groups of people (for instance, forestry officials and local communities) is the way in which they are connected. It is the former, which is of concern here: the relationship between various stakeholders and the forest. The relations between different groups of stakeholders are also extremely important, but will be discussed later in relation to the balances and imbalances of each group of stakeholders' rights, responsibilities, returns from forest resources and relationships.

In the North, it is widely believed that, since Africans lack an emotional experience with romanticism and transcendentalism, they do not possess

[1] This section derives mainly from Burnett & Kang'ethe (1994).

the philosophical prerequisites necessary to protect wilderness. However, the North's disdain for African systems of thought has precluded examination of customary African views of wilderness and forest (Burnett & Kang'ethe 1994). Examining relationships to forest is a likely place to begin to understand African philosophies of natural forest management and conservation.

Northern philosophies of wilderness have been dominated by anthropocentric and biocentric theories (Vest 1991). The former theory allows humans to control wilderness absolutely for egocentric purposes while the latter gives nature absolute control, but without aesthetic purpose. In either case the Northern perspective results in wilderness filled with parks, forest reserves and game reserves.

Customary East African religion is fundamentally monotheism (Burnett & Kang'ethe 1994). It is believed that everything in the universe has vital energy or being and interaction strengthens this being. The power to strengthen, regenerate and heal is measured in a hierarchy of five ontological categories: God, ancestors or spiritual beings, humankind, plants and animals, and non-biological things (Mbiti 1970).

Certain places are more associated with God than others. Among these are high places and mountaintops (Burnett & Kang'ethe 1994) that are also wilderness (Burnett & Kang'ethe 1994; Feierman 1974). Healing and regenerative powers thrive in wilderness, but not in the domesticated sphere (Feierman 1974). Burnett & Kang'ethe (1994) believe that high elevations as wilderness were sacred by virtue of the fact that tropical crops were unadapted to the cold and could not grow there. However, Kajembe (1994) and Feierman (1974) believe wilderness areas to be sacred primarily due to an association with water or rain.

The attitude to wildlife was hostile and while it could generally be killed without a second thought, the clearing of forest presented innumerable problems (Burnett & Kang'ethe 1994). Clearing of forest for agricultural land, and hence domesticating the wilderness, could not take place without taking pains to relieve the spiritual dilemma created (Routledge & Routledge 1910). The usual explanation is that trees were associated with spiritual beings, hence the largest species of trees were retained to enable the spirits to accumulate and find safe abodes. When age and events overtook even these trees, special precautions in the form of rituals were required to assure that the spirits could find new homes (Bildsten 1998: personal communication; Burnett & Kang'ethe 1994).

Certain trees maintained a special association with God. Burnett & Kang'ethe (1994), Fleuret (1980), Johansson (1991), Kajembe (1994) and Woodcock (1995) have noted the genus *Ficus* to be particularly common as a sacred tree. Not only was the tree sacred, but so was the land around it and everything it sheltered. A group dared to approach God about matters and events particularly troublesome to a group only at the sacred tree. The sacred tree was never wilderness and that rather than being an isolated, distant object, it was a near neighbour.

In the light of customary African attitudes to wilderness and more specifically forest, the anthropocentric and biocentric theories of the North, whose perspective of wilderness results in forest reserves, contrasts dramatically with the African. Northern solutions to forest management fall into the African ontological categories four and five: plant and animal, and non-biological. Burnett & Kang'ethe (1994) argue that forest management systems created on these arbitrary categories are unacceptable and that customary African relationships to the forest should be examined in order to employ an African philosophy of conservation and natural forest management (Chapter Four).

Rights and Responsibilities to Forest[2]

Rights and responsibilities in natural forest management are discussed jointly, since in combination they are embodied in tenure regimes. African tenure regimes are defined by Shepherd et al (1995: cited in Dubois 1997: 3) as *socially defined rules for access to resources and rules for resource use that define people's rights and responsibilities in relation to resources*. They reflect relations between different stakeholders. This is in contrast to tenure regimes outside rural areas of Africa, where security is often equated only with ownership, hence control. Dubois (1997) argues that these contradictory views have led to misunderstanding and inefficiency in land and forest management.

It is important to clarify the concept of tenure security, as different stakeholders interpret it in different ways. Government and international bodies usually associate tenure security with its spatial aspects, whereas the customary view relates it more to securing social relationships (Dubois 1997). Feierman (1990: 11) suggests that there is a general structural relationship between 'ownership' and the benevolent use of ritual power:

> Powerful medicinal substances cannot be used to heal unless they also have the capacity to kill. The most certain guarantee that the possessor of medicine will use his power on the side of life rather than death is to give him, as his own property, all that must be preserved. For example, in the rite of sacrifice, members of the sacrificing lineage all left their house at a

[2] This section derives mainly from Dubois (1997).

certain point and then re-entered, explaining in a song that the house now belonged to the visiting healer, a dangerous practitioner. He would use his medicines to sustain the house because it had become his own.

Similarly, in order that leaders healed the land (**kuzifya shi**) by bringing rain, rather than harmed the land (**kubana shi**), the wealth of the land in Usambara had to be given to the king through tribute (Feierman 1990). 'Ownership' or tenure security of the land has a strong political sense. *Any policy, which attempts to increase tenure security in Africa, should develop mechanisms that not only validate tenurial rights according to formal law, but also allow for social validation and clarification as to their socio-political implications* (Chauveau 1996: cited in Dubois 1997: 3).

A growing body of evidence shows that the relationship between tenure security and a more efficient and sustainable use of the resources is not always straightforward (Dubois 1994; Platteau 1996). In fact, many observers agree that usufruct or management rights may often be more important than ownership, so long as confidence in future access and mutual recognition exist (Borrini-Feyerabend 1996; Dubois 1994; Platteau 1996).

A pragmatic way to look at property rights is through a continuum approach, whereby relative access to the resource distinguishes private from common property, as underlined by Messerschmidt (1993: cited in Dubois 1997: 3):

> If we view private and common property on a continuum, private property falls at one end, defined by the most delimited rules of access and the most restrictive rights of use. Common property takes up the rest of the continuum, with the commons of a whole community at the far end and communal commons and government reserve falling somewhere in between.

An advantage of the continuum approach is that it accommodates the notion that common property regimes (CPRs) can encompass shared private property (McKean & Ostrom 1995). Under such systems, resources are not divided into pieces. Holders of such rights have to abide by some obligations, common to all holders of land in the communal area.

There is intense debate over the pros and cons of CPRs. To provide some clarification, myths and realities surrounding CPR are summarised in Table 1.6.

The CPR approach recognises four categories of property rights: private, common, state, and open access (for instance, no property rights). Dubois (1997) in a commentary on Lynch (1992), accepts that this categorisation

Table 1.6 Some myths and realities about common property regimes

Myth	Reality
Individual gain provides a stronger motivation than communal good and, as a result, CPRs are over exploited.	Individual survival and security, both in terms of material resources and social identity, is dependent on community survival, particularly within a harsh environment.
As a resource becomes more valuable and/or there is increased pressure to use that resource, over exploitation and a cycle leading to degradation is inevitable.	If increased pressure on CPRs comes from within the society of the original resource users; they will evolve responses to manage it. However, if it comes from more powerful outsiders, CPR organisational rules will be ignored and/or will need support to adapt on time.
CPRs are 'impure' public goods (where one person's use subtracts from the use of others). This subtractability works in two ways: • firstly, any user of the commons subtracts from a flow of benefits to another; and • secondly, cumulative use by increasing numbers will eventually lead to a reduction in the productive capacity of the resource.	In reality, CPR management decisions are constantly taken and enforced. 'Free riders' are quickly sanctioned. It is important to acknowledge the interdependence of communities, which manage CPRs.
Common property is an inefficient way to manage resources.	Common property management is the most efficient way of managing certain resources; for example rangelands with limited, seasonal water sources.
Such systems rarely maximise production.	Maximising production is not always the user's main aim: reducing, or spreading risks is often more important, together with the presence of a safety net in times of hardship.

Source: After Shepherd et al 1995.

contains two major flaws:

- It creates confusion between private and individual ownership; and
- It requires disentangling individual and group rights for recognition in community-based systems.

He therefore provides a simplified way to analyse tenure, distinguishing only two types of rights (for instance, 'public' or state owned and 'private', referring to rights held by non-state entities, whether individuals or groups). This leads to four possible combinations (Table 1.7).

Table 1.7 Land tenure regimes

	Individual	Group
Private	For example, titled land.	For example, ancestral domains.
Public	For example, leases on individual farming or woodlot parcels.	For example, communal leases.

Source: After Dubois 1997.

Although rural dwellers usually see trees as part of their farm land, it is often convenient to separate land tenure from tree tenure, as it allows for a better perception of the formal reality.

Tree tenure The concept of tree tenure, as opposed to land tenure, has gained momentum in the last decade. This stems from the increasingly acknowledged observation that the relation between these two concepts is usually complex, given that rights to trees can be multiple and separable from land.

Fortmann (1985: cited in Dubois 1997: 5) defines tree tenure as *a bundle of rights, which may be held by different people at different times.* She distinguishes four major categories of tree tenure rights:

- The right to own or inherit;
- The right to plant;
- The right to use; and
- The right of disposal.

The right of disposal encompasses notably the rights for stakeholders - especially local communities - to decide that forest resources will not be commercialised. This right is often overlooked in the analysis of community forestry activities, despite its importance (Ribot 1995).

Fortmann (1985) discerns three sets of factors that influence rights over trees:

- The nature of the tree: Tree tenure regimes are very variable. They may differentiate planted trees from wild trees, or trees on private land from trees on common land. Distinction is often based on the principle that 'labour creates rights'. In general, wild or self-sown trees on common land are common property. Planted trees belong either to the planter or to the owner of the land. In some cases, wild trees growing on private land belong to the landowner.

- The nature of the use: The distinction is often made between subsistence and commercial uses. Trees used for subsistence are often free for use by all, especially when they grow on communal areas. In the case of commercial use, the right may be restricted to trees growing on the seller's property; it may also be forbidden, depending on the use. Fuelwood can fall into both categories and, where it has become scarce, rules governing its use tend to be tightened up.
- The nature of the land tenure system: Tree and land tenure affect each other. Where land tenure is communal and tree tenure is strong, the planter of a tree generally owns that tree. This very often characterises places where shifting cultivation is the main agricultural system. In contrast, where private rights to land are strong, rights to trees are more correlated to land rights. This favours landowners, at the expense of groups with weaker land rights, for example, tenants, squatters, mortgagees, and women. These groups have restricted rights due to the possibility of using tree planting to claim permanent rights on land.

Land tenure Following Hesseling and Ba (1994: cited in Dubois 1997: 7), land tenure refers to relations among individuals and groups that govern the appropriation and use of land. In most rural areas of sub-Saharan Africa, there is a dual system of tenure: the formal and the customary.

The dynamic character of 'current' customary rules is emphasised by Bruce (1993) who argues that, nowadays; important customary rules often turn out to be only a generation old. Dubois (1997) therefore finds it more appropriate to use the term indigenous rather than customary when referring to locally drawn rules. This implies, however, that 'past' customary rules were static. Calling them 'indigenous' rules also implies that they are locally drawn, when in fact many current customary rules have evolved and developed due to and from outside sources, particularly under colonial administrations (for example, Conte 1996).

The dual tenure system is often aggravated by the lack of means for the state to exert its statutory role. Too much control, without adequate means to enforce regulations ends up with the state actually losing control and the resource becoming in fact open access. As a result local people tend to resort to developing new customary rules to utilise the natural resources under the new conditions. With a lack of customary management rules, which have evolved over time and experience, due to their curtailment under colonial administrations (Borrini-Feyerabend 1996), the new customary rules tend to be to gain land tenure through deforestation or forest degradation (Dubois 1997).

This lose situation in terms of rights over land and its resources, further compounded by the inadequate wages paid to forestry sector staff, has often resulted in the prevalence of covert arrangements between stakeholders at the local level, for example, replacement of official fines by bribes and clientelism. The result of bribery and rent-seeking is, however, that the individuals will seek to access the resource as quickly as possible, so as to derive the best and most out of it before leaner times, changes in authority and so on.

A variety of solutions have been put forward to resolving these issues of insecure land tenure. One such solution is to move towards more formalisation of private individual land rights. Titling is intended to bring more security and higher prices for land, leading to better management of natural resources. However, the World Bank itself admits that nearly all its rural titling schemes have achieved 'poor' results (*The Economist* 1995). As regards forests, titling is virtually impossible for individual properties, and protection from outside encroachment has to be sought across groups of people (Hobley 1995).

Given the weak enforcement of state regulations, more and more attention has been given to customary rules as possible alternatives to regulating the use and management of natural resources in rural areas of Africa. It is sometimes argued that the codification of customary rules would help their recognition at government level.

However, as Dubois (1997) points out, this task would be extremely difficult given the complexity, localisation and dynamics of many customary rules. Formalisation through the enactment of legislation to regulate customary rules is sometimes claimed as another possible mechanism to avoid dualistic tenure regimes (Dubois 1997). However, there are several disadvantages (Mortimore 1996: cited in Dubois 1997):

- The government assumes the right to define the goals and is not an independent party with respect to natural resources;
- The complexity and flexibility of customary rules are almost certain to be overlooked; and
- Organising tenure along formal principles and using formal tools (for instance, titles) is very expensive and time consuming.

Dubois (1997) feels that a more promising approach would be to develop mechanisms that officially recognise some customary rules, but on an *ad hoc* basis (for example, see Raharimala 1996 & Ranzanaka 1996: both cited in Dubois 1997).

In Tanzania, land and tree tenure have been studied to varying degrees (for example, see Dobson 1940; Johansson 1991; Kajembe & Mwaseba 1994; Kessey & O'Kting'ati 1994; The United Republic of Tanzania 1994). However, with reference to natural forest management specifically, only Wily (1997 & 1999) has investigated rights and responsibilities and this has been in the Miombo woodlands, rather than in high biodiversity forests, such as the Eastern Arc, upon which this research is based.

Returns from Forest Resources

'Returns' from forest resources refer here to the profit, income, earnings, gain or harvest obtained from forest resources. The literature tends to divide these returns into two main categories: forest products and forest services. Forest products tend to also be sub divided into timber and non-timber forest products (NTFP). The definition of NTFPs used in this thesis is that used by de Beer & McDermott (1996): *The term non-timber forest products encompasses all biological materials other than timber which are extracted from natural forests for human use.* Forest services include both ecological services and forest land.

In the early 1990s much research in Tanzania was initiated into NTFPs and indigenous knowledge, with the onset of participatory approaches to biodiversity conservation and natural forest management. Past conservation policies were often associated with restrictions and loss of resources/returns by local people. However, in the participatory era, there was an increasing awareness that conservation and successful management of forest could not be achieved without the co-operation and support of local peoples (Wells & Brandon 1993: cited in IUCN 1997). The amount of research into NTFPs can be demonstrated by the fact that there are a number of bibliographies published on NTFPs (for example, IUCN 1997; Scoones et al 1992).

Much of the research in Tanzania initially concentrated on inventories of NTFPs (for example, Cambridge-Tanzania Rainforest Project 1994; Härkönen et al 1995; Matthews 1994; Mogaka 1991; Owen 1992; Ruffo 1989; Ruffo et al 1989; Warburg 1894) with others looking further at local dependence on NTFPs (for example, Emerton 1996; Evers 1994; Fleuret 1979a; 1979b & 1980; Lagerstedt 1994; Lindstrom et al 1991; Woodcock 1995).

Assigning economic values to forests beyond timber exploitation is regarded by some (in terms of NTFPs, see for example, de Beer &

McDermott 1996 and in terms of land, see Norton-Griffiths & Southey 1995) as a powerful argument to enhance sustainable use incentives and to convince policy makers that the environmental functions and products from tropical forests are important at local and regional levels, as well as nationally. Wily (1997: 3) is however, scornful of what she calls *State of the Art Calculations of Local Forest Values,* arguing that it is the socio-political issues that are more important than the socio-economic.

While efforts to collect information on NTFP utilisation increased, little research has been done to assess whether the harvesting methods of NTFPs are sustainable (IUCN 1997). It is often assumed that since local people have been collecting and utilising a variety of NTFPs for extended periods of time, their harvesting methods are sustainable. However, population growth, diminishing forest habitat, outside interventions and external influences are thought to be affecting this (IUCN 1997). Wily (1997) believes however, that with local management rights, local people would be the best judges of whether harvesting rates are sustainable or not. For instance, when Mpanga community, in East Usambara, made management plans for their community forest, they chose to close the forest from collection of NTFPs (Bildsten 1998: personal communication).

Research into returns from forest resources: NTFPs, timber, forest land and indigenous knowledge; is an important issue for biodiversity conservation and natural forest management (Chapter Three). However, it is only one aspect of stakeholders' roles in an approach where local people are managers of forest resources not just beneficiaries.

Summary

As the literature demonstrates, approaches to biodiversity conservation and natural forest management have evolved in Africa over recent decades. These changes in approach affect various stakeholders' in forest resources. Changes in approach, in turn lead to changes in stakeholders' respective roles. Defining stakeholders' roles via their respective rights, responsibilities, returns from forest resources, and relationship to forest, their '4Rs', highlights any imbalances between these attributes, which may in turn impair adequate negotiation and lead to deforestation and forest degradation.

Approaches to attempt to rebalance these attributes have evolved noticeably in South Asia, through Joint Forest Management and Community Forest Management. However, in comparison to experiences in South Asia, community involvement in natural forest management in Africa has been slow to evolve and has been largely confined to discrete, usually donor-prompted, initiatives. In Tanzania, there have been two main cases of community forest management, but these have been in the Miombo woodlands rather than in forests of high biodiversity, where conservation priorities are high, such as the Eastern Arc forests. However, the Tanzanian National Forest Policy, announced in 1998, gives hope for more meaningful community involvement in forest management through Joint and Community Forest Management.

Research into stakeholders' roles in natural forest management in Tanzania has focused primarily on the returns from forest resources, with only one author (Wily 1997) looking closely at rights and responsibilities and the relations between stakeholders. In the Eastern Arc forests only returns from forest resources have so far been studied. Nothing has been written specifically about the relationship between stakeholders and forests in Tanzania, although some authors (for example, Johansson 1991) have examined the relationship between stakeholders and trees. Little has also been written about the changes in these imbalances through evolving management approaches.

In an attempt to examine the changes in stakeholders' roles and the imbalances in their '4Rs', through changing forest management approaches in the Eastern Arc Mountains, the following Chapters hope to contribute to the development of improved mechanisms for the negotiation of roles in natural forest management. Understanding the limitations of previous management may give a better understanding of the opportunities for present and future management approaches, which may be conducive to sustainable forest management.

Chapter 2

The Eastern Arc Mountains

Forests of the Eastern Arc Mountains

The Eastern Arc forests are situated on the chain of Precambrian block, faulted mountains (Griffiths 1993) - the Eastern Arc Mountains - that cut across Tanzania (Figure 2.1). The Arc comprises 11 separate mountain blocks, starting from the northeast: the Taita Hills in Kenya, the North and South Pare, East and West Usambara, Nguru, Ukaguru, Rubeho, Uluguru to the Udzungwa and Mahenge Mountains in the south (Figure 2.1).

The forests are thought to have formed 100 million years ago and been isolated from each other for the last 20 million years and represent one of the oldest and most stable ecosystems on the African continent (Sayer et al 1992). During times of climatic change these forest islands have remained stable due to their ability to trap the humid air of the Indian Ocean, thus having a higher rainfall and so maintaining their forests. This stability has enabled the Arc to evolve a highly specialised flora and fauna. The mountains are sometimes called the 'Galapagos Islands of Africa' due to their small and fragmented island forests and high levels of endemism (Lovett & Wasser 1993).

In total the Arc covers less than two percent of Tanzania's land area, but harbours 30 to 40 percent of the countries species of flora and fauna (Hamilton & Bernsted-Smith 1989). Over 60 percent of all Tanzanian endemic plant species are in the Eastern Arc and over 25 percent of Eastern Arc plant species are endemic (Rodgers 1997). There are 16 endemic plant genera in the Eastern Arc Mountains. Several genera of herbs and shrubs have undergone remarkable radiation in the Eastern Arc Mountains, often with many of the total world species of highly restricted endemism, in the Arc. For example, the African Violet (*Saintpaulia spp.*), which has 20 of the 21 species in the world endemic to the Arc (Rodgers 1993). The Arc does not only support wildlife but forms an essential foundation for the country's livelihood through ecological services to the people and communities of the mountains and coast.

Moreau (1935) classifies the forests into three main types based on

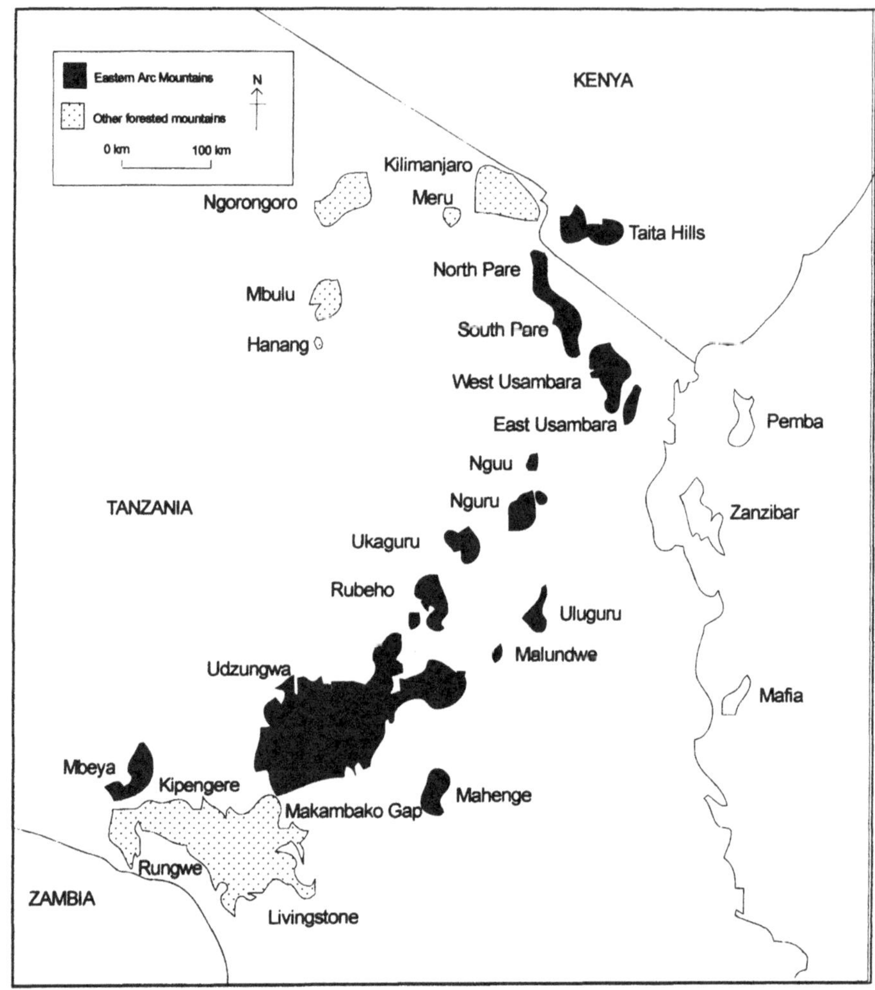

Source: Modified from Lovett & Wasser 1993.

Figure 2.1 The Eastern Arc Mountains

structure, climate, dominant species and avifaunas:

- Lowland Forest, below 800 metres;
- Submontane Forest, between 800 metres and 1,500 metres; and
- Montane Forest, 1,500 metres and above.

This research focuses on the forests of two blocks of the Eastern Arc

Mountains in Tanzania: the East Usambara Mountains and the Udzungwa Mountains (Figure 2.1). Table 2.1 lists the case study forests by name, mountain block, forest type, forest status, forested land area and forest-local communities. Case study forests were selected by ensuring that as a group they fit the following criteria:

- Eastern Arc forest;
- Range of Lowland, Sub-montane and Montane forest types;
- Full range of legal forest status: Central Government Forest Reserve (CGFR), Local Government Forest Reserve (LGFR), Public Forest and Private Forest; and
- Forestry related Non-Governmental Organisation (NGO) activity in forest-local communities.

Figures 2.2 and 2.3 show the location of East Usambara and Udzungwa case study forests respectively.

The East Usambara Mountains

The East Usambara Mountains (Figure 2.1) start 40 kilometres from the coast and rise sharply to over 1,000 metres and peak at 1,500 metres. Rainfall distribution is bi-modal, peaking between March and May (***mwaka***: long rains) and between September and December (***vuli***: short rains). The dry seasons are from June to August and January to March. Rainfall increases with altitude from 1,200 millimetres annually in the lowlands to over 2,200 millimetres in the highest altitudes (Hamilton & Bernsted-Smith 1989). Totals are higher on southeast facing slopes as they are exposed to the moist winds from the Indian Ocean. The lowland mean temperatures are typical for their altitude, in contrast to the abnormally cool climate of the uplands. Moreau (1935) gives a mean temperature of 24°centigrade and mean annual rainfall of 1,650 millimetres for a site at 200 metres in the Middle Sigi Valley.

Due to their age, isolation and their function as condensers of the moisture from the Indian Ocean, they support ancient and unique forests, rich in endemic species and high in biodiversity (Hamilton 1988; Howell 1989; Rogers & Homewood 1982). Currently, around 2800 taxa of plants have been recorded of which it is suggested that over a quarter are endemic or near endemic (Iversen 1991). Many are threatened (Rodgers 1996).

Research in the East Usambara Mountains began in the late 1890s with substantial botanical collections being taken. In 1928 amphibian surveys

Table 2.1 Case study forests and forest-local communities

Forest Name	Mountain Block	Forest Type	Forest Status[3]	Forest land area (ha)	Forest-local communities
Kambai	East Usambara	Lowland	CGFR: (Proposed 1974; Gazetted 1992; Extended 1993).	1046	Kambai & Kwezitu
Semdoe	East Usambara	Lowland	CGFR: (Proposed 1964; Gazetted 1993; Extended 1998).	1000	Kambai & Seluka
Manga	East Usambara	Lowland	CGFR: (GN 112, 1955).	867	Mkwajuni & Kwatango
Mlinga	East Usambara	Sub - montane	CGFR: (Proposed 1954; Gazetted 1994).	190	Miembeni, Gare & Magula
Lulanda	Udzungwa	Montane	LGFR: (Boundary marked in 1941; not known to be officially gazetted).	196.7	Lulanda
Kambai	East Usambara	Lowland	Public Forest	10	Kambai
Kwezitu	East Usambara	Sub - montane	Public Forest	100-200	Kwezitu
Magoroto Hill	East Usambara	Sub-montane	Public Forest	10	Mgambo, Miembeni, Gare
Magrotto Estate	East Usambara	Sub - montane	Private Forest: (GN 99, 1931).	200-300	Mgambo & Miembeni

Source: Forest status and area data taken from EUCFP 1995.

were undertaken and ornithological work began. Biological research in East Usambara has steadily increased over the years, with the Frontier-

[3] Where CGFR is a Central Government Forest Reserve and LGFR is a Local Government Forest Reserve.

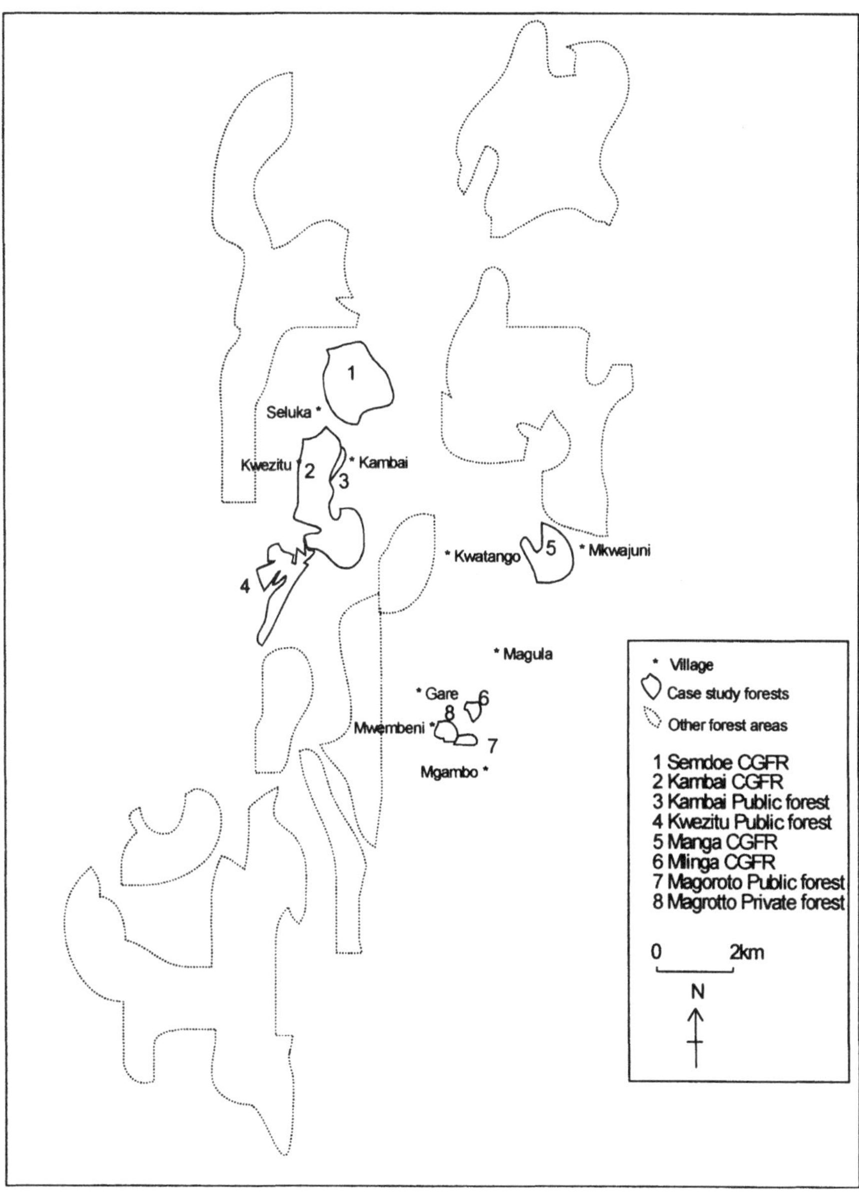

Source: Modified from Johansson 1994.

Figure 2.2 East Usambara case study forests and forest-local communities

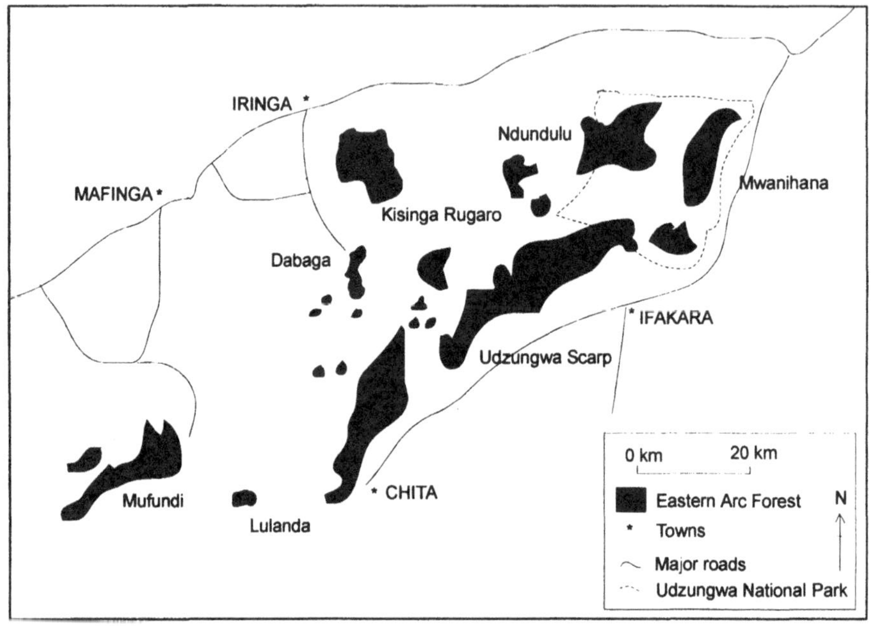

Source: Modified from Rodgers & Homewood 1982 (with permission from Elsevier Science).

Figure 2.3 The location of Lulanda forest in the Udzungwa Mountains

Tanzania Biodiversity Surveys undertaken since 1994.

Additional to the biodiversity value is the drainage and catchment value of the East Usambara Mountains. The forests play an important role in maintaining the regional hydrological cycle, which feeds the Sigi River. The Sigi River is a vital water source for the local communities as well as supplying water for the large coastal town of Tanga. Mount Mlinga, at 1069 metres, is the source of the Mkulumuzi and Mruka rivers, which drain Magoroto Hill. The Muzi, Semdoe, and Miembeni drain the mountains bordering the Sigi Valley, which flow into the Sigi River.

The latest survey of the area, conducted by Johansson and Sandy (1996) shows that approximately 45,137 hectares of East Usambara remain as forest. This can be divided into two types: submontane and lowland forest (Moreau 1935). Altitude is the factor differentiating these two types with submontane forest generally occurring above 800 metres. Submontane forest occupies 12,916.6 hectares (30.7 percent); lowland forest occupies 29,497.4 hectares (62.9 percent) and forest plantation 2,723.6 hectares (6.5

percent); 21,900 hectares are presently gazetted forest reserves. The remainder, 35,909 hectares (43 percent) is classified as agricultural land; woodland; grassland; ponds; rivers; barren land; and settlements (Johansson & Sandy 1996).

The geology is predominantly acidic Precambrian basement rocks (Iversen 1991). The montane soils are highly leached and acidic with little fertility in contrast to the lowland soils, which have more bases and are more productive in terms of agriculture. The boundary between typical laterized (ferralsols) and non-laterized soils (ferrisols) has been put at 900 metres on the wetter east and southeast sides (Iversen 1991). Soils are generally deep - between one and five metres - red or bright sandy clays or sandy loams (Cambridge-Tanzania Rainforest Project 1994).

The Udzungwa Mountains

The Udzungwa Mountains are an extensive block mountain range covering some 10,000 square kilometres and running from northeast to southwest (Figure 2.1). They rise from a level of 300 metres at the valley of the Great Ruaha River in the north as a series of rolling hills and dissected plateau to a gently undulating upland (1,200 metres) with peaks reaching 2,800 metres, and end as a steep south-east facing scarp.

The Udzungwa Mountains experience the single rainy season of southern Tanzania, with the bulk of the precipitation due to the November to May southeast monsoons. Rainfall is high (1,800-2,000 millimetres per year) along the southeast scarp but decreases rapidly down the northwestern rain shadow slopes towards Iringa (660 millimetres per year). The wet southeast slopes receive some rain throughout the year (only three months with less than 50 millimetres) while the dry northern areas suffer a six months' drought. Water runoff and stream density reflect this rainfall pattern, the south-eastern slopes having permanent fast flowing streams that run off into alluvial fans at the scarp foot. Permanent streams are much fewer to the northwest. Soils of the higher areas are lateritic, slightly acid red earths of medium fertility.

In the past the bulk of the mountain block is thought to have been covered in forest, with lowland and montane rain forest along the wetter slopes; dry evergreen forest on the plateau and deciduous forest grading to woodland and thicket on the drier northern slopes (Rodgers & Homewood 1982).

Political History (post 1740[4])

What is now known as Tanzania, was in the eighteenth century a series of separately ruled kingdoms and chiefdoms (Kimambo 1996). German colonial interests were first advanced in 1884. Karl Peters, who formed the Society for German Colonisation, concluded a series of treaties by which tribal chiefs in the interior accepted German 'protection'. Prince Otto von Bismarck's government backed Peters in the subsequent establishment of the German East Africa Company. In 1886 and 1890, Anglo-German agreements were negotiated that delineated the British and German spheres of influence in the interior of East Africa and along the coastal strip previously claimed by the Omani sultan of Zanzibar. In 1891, Tanganyika became German East Africa, as the German Government took over direct administration of the territory from the German East Africa Company and appointed a governor with headquarters at Dar es Salaam.

German colonial domination of Tanganyika ended in 1918, after the First World War, when control of most of the territory passed to the United Kingdom under a League of Nations mandate. After the Second World War, Tanganyika became a United Nations trust territory under British control. Tanganyika became independent in 1961 and united with Zanzibar to form the United Republic of Tanzania in 1964.

The East Usambara Mountains

The largest and traditionally rivalling tribes in Usambara were the Zigua, Bondei and Sambaa. Prior to political centralisation in Usambara by the Kilindi around 1740, farming communities lived in locally centralised neighbourhoods under the auspices of certain lineage heads or clan leaders (Conte 1996). The only known Usambara king prior to the Kilindi dynasty was king Tuli, ruling in Vugha in West Usambara (Iversen 1991). The arrival and take over, around 1740 of Mbegha, the first Kilindi ruler of the Sambaa, is shrouded in myth. Mbegha is said to have come from the Nguu Hills to the south of Usambara and his kingdom covered West Usambara. In 1790 Mbegha's son, known as Shebughe became king in Vugha and peaceful development continued (Iversen 1991).

In 1800, Shebughe expanded the kingdom southwards into Zigua country and conquered East Usambara. Mbegha's second son Maua

---

[4] The political history described here is post 1740 since it is thought to be the approximate date of the Kilindi take over in Usambara and the initiation of political centralisation in the area. It should be noted that little has been written on precolonial history in Udzungwa.

reigned for a short time in the 1800s. According to Feierman (1974), Maua is virtually always excluded from the traditions, due primarily to a short, weak, unpopular reign. However, key informants in Mgambo village, Magoroto Hill in East Usambara mention Maua as being the first Kilindi ruler in the area (Author's fieldwork 1994-1998). Then another of Mbegha's sons, Kimweri ye Nyumbai became king in 1815 and continued the expansion, holding the whole coastline below Usambara by 1835. Informants from Magoroto Hill told how many inhabitants came to the area from the lowlands to escape famine or from the West Usambara Mountains to escape tribal wars in the early 1800s (Author's fieldwork 1994-1998). The majority of informants talk of the first rulers being sons sent by Kimweri ye Nyumbai to rule in East Usambara. Feierman (1974) believes that this popular tale means, that with the coming of Kimweri's sons, all previous chiefs became irrelevant.

Kimweri's son, Semboja was the biggest slave trader in the Usambara history (Feierman 1974) and ruled from Vugha, whilst another of his sons Mnkande ruled East Usambara. Mnkande placed two of his sons, Makange and Kibanga, in East Usambara and one Shekulwavu in the north of West Usambara. Makange ruled the north of East Usambara whilst Kibanga made his capital in Mghambo in East Usambara after his brother Chanyeghea died in 1870. Mghambo was on a mountain peak overlooking the valley to the west. It was extremely difficult to reach: a useless position for trade but easy to fortify and defend against invaders who would have to climb the mountains (Johnston 1879: 550-51):

> The town was encircled by impenetrable brush, and outside this there was a second barrier of felled trees and deep trenches. The gateways were in walls of posts driven into the ground, so as to form a mass six feet thick. Each gate had two doors formed of heavy single slabs of timber.

Kinyashi (Shekulwavu's son) conquered East Usambara, fully supplying the capital with tribute, but it also added a great body of subjects who could not have been loyal, as they were not considered true Sambaa subjects. Two generations after Kimweri, East Usambara subjects, led by Makange, Kibanga and the Bondei destroyed the capital (Feierman 1974).

The German take-over in Usambara was violent. A revolt started in 1888 in Pangani and led by Abushiri bini Salim and Bwana Heri was soon spreading in the area (Förster 1890: cited in Iversen 1991). However, the German troops were supported by British and Portuguese forces, and Abushiri was betrayed and hanged in December 1889. Bwana Heri capitulated, but tried, unsuccessfully, to rise again in March 1894 (Iversen

1991).

The nominal king in Vugha, Kimweri the Second Maguvu, died in 1893 and was succeeded by his brother Mputa (Doring 1899: cited in Feierman 1974). As Mputa did not want to co-operate with the Germans, he was hanged in Mazinde in 1895, and later the same year even Semboja died (Feierman 1974). For a short period the Germans installed Kipanga from Handei as king in Vugha, but as he was not a Kilindi, he was soon replaced by the right heir Kinyashi, a great grandson of Kimweri the First (Doring 1899). Kinyashi abdicated in 1903, afraid of being assassinated by his own people. The system of royal village chiefs from the Kilindi clan, nominally continued during the first years of German rule, but often political agitators (***akidas***) sent from the coast by the Germans to 'accompany' local chiefs held the real power (Feierman 1974).

In the British administration, the German estates in Usambara were confiscated, but in 1925 the original owners were given permission to return to their estates, and about half of them did so (Hamilton & Bernsted-Smith 1989). The last German estate owners left the area at the end of the Second World War.

A Native Authority Ordinance was issued in 1926, giving the native authorities control of burning practice and water use (Watson 1972). The Kilindi dynasty was nominally re-instated in Usambara (Cliffe et al 1975). Kinyashi, having abdicated in 1903, was induced to return in 1926, but he resigned again in 1929 when Billa Kimweri, a grandson of Semboja, was elected king (Winans 1962). Billa Kimweri died the same year and was succeeded by his brother Magogo. In 1947, 5000 people demonstrated to remove Magogo, resulting in the take-over of his son Mputa (Egger & Glaeser 1975: cited in Iversen 1991). The king and the elected Kilindi village chiefs had not much more than nominal power and they were often quite unpopular among the native farmers. After spreading local opposition, the whole system was abandoned in the 1950s (Cliffe et al 1975), and after independence all traces of the local Kilindi rule were replaced by a system of party (TANU, later CCM) chiefs and offices.

The Udzungwa Mountains

The inhabitants of Udzungwa, the Hehe, lived both in the highlands and lowlands. Redmayne (1968), writing about the Hehe before 1900 remarked that it was difficult to write a detailed history of the Hehe, because first and foremost, even the name itself was only recent, and it was not used to denote a specific ethnic group. He further states that the name 'Hehe' was mentioned for the first time in writing by Richard F. Burton an English explorer who in 1857 travelled across Ludi, north of Lya Mbangali (Ruaha Kubwa). Burton believed then that the area he had crossed was the central area of Uhehe (land of the Hehe people). Redmayne (1968) believes that the name 'Hehe' originated from the 'war-cry' when fighting in the battlefield. They would shout "Hee! Hee! Hee! Kill the enemies, ehee!" (Redmayne 1968). From this Redmayne (1968) believes they came to be referred to by others as the 'Hee-hee' people – the Hehe.

The most well known chief (**mutwa**) of the Hehe was Mkwawa, full name Mukwav'inyika (Madumulla 1995). Mkwawa who was born in 1855, owed his position partly to his genealogical position as a member of the Muyinga clan, partly to his own ability and intelligence and partly to supernatural sanctions (Madumulla 1995). He was believed to have a special relationship with the spirits (**masoka**) of the dead chiefs to whom he was related and made offerings on their graves to ask for assistance in matters which concerned the whole chiefdom (Madumulla 1995). In 1898 Mkwawa took his own life so as to avoid the act of being captured alive by the German colonialists against whose invasion and rule he had fought for seven years, in both face-to-face and guerrilla warfare.

Evolution of Approaches to Natural Forest Management in the Eastern Arc Mountains

Pælaeoecologists and prehistorians argue that forest cover was at a maximum spread some 6000-8000 years before present and has decreased partly due to a gradually drying climate and the effects of burning by hunter-gatherer and agricultural communities (Rodgers 1993). In East Usambara, Schmidt (1990: cited in Rodgers 1993) presents evidence of archaeological finds in the surface soil layers of submontane natural forest sites which are thought to have derived from early Iron Age settlements some 2000 years ago. Schmidt (1990) argues that this was part of a settlement pattern invading forests all over East Africa between 500 BC and 500 AD. He describes a later wave of village settlements and clearings in the later Iron Age, at about 900 and 1000 AD.

Conte (1996), Feierman (1974) and Kimambo (1996) have examined respectively the ecology, politics and economics of land use in Usambara post 1740. The Sambaa generally preferred to live on the high ground at altitudes of approximately 1,000 metres, which they called **Shambaai**[5]

[5] This zone was called **Shambaai**, since in Kisambaa the addition of the final 'i' creates the

(Feierman 1974; Kimambo 1996). Many explanations have been given for this preference. Kimambo (1996) and Conte (1996) argue that in contrast to the surrounding **nyika** (lowland) there was more rain, more water and there were therefore better possibilities for producing food on the mountains. The Sambaa also recognised the health benefits of living in the mountains, where there was less chance of becoming sick with malaria (Kimambo 1996). Fleuret (1978) argues that the Sambaa congregated in villages to defend themselves against the Masai, Taita and other neighbouring people. Feierman (1974) stresses that compact small chiefdoms facilitated Kilindi collection of tribute in labour or kind, but Kimambo (1996) argues that the Kilindi found the pattern already established and exploited it. He goes on to say that by settling in the **Shambaai** zone, they got to use both the **nyika** (lowland) and the cool areas at higher altitudes. Therefore, he concludes, it was the pursuit of ecological and economic advantages rather than political control that located the Sambaa in the submontane forests of Usambara.

Civil war between 1855 and 1895 and the region's increasing importance as a source of slaves interrupted the intensification of Sambaa agriculture, which had been occurring since the early nineteenth century. Slave raiding forced the Sambaa people to live in heavily fortified hilltop communities. Women who had previously cultivated their maize fields several miles from their villages either farmed closer to home on overused, less fertile fields, or were escorted to work by men who neglected their irrigated fields (Fleuret 1978). Holst (1893: cited in Iversen 1991) described in detail the impressive irrigation system, but also observed large areas of forest on steep slopes converted to farmland. Traditional shifting cultivation was continuing way up to the hilltops. Forest did not regenerate where slash and burn had been practised (Buchwald 1897: cited in Iversen 1991) and large forests broken up by farming and bush fires were common in drier areas (Baumann 1889 & Doring 1899: both cited in Iversen 1991). The neighbourhood of Vugha was completely denuded early on, and firewood had to be collected from long distances (Krapf 1858: cited in Iversen 1991; Meyer & Baumann 1888). The agricultural system designed to ensure food security was no longer working properly, because unrest interfered with local customary land management approaches.

These accounts of deforestation and forest degradation came from West Usambara. The early visitors have given the general impression that East

---

locative form. The Shambaa are the people and *Shambaai* is their home (Feierman 1974). Note that Feierman (1974) uses the pronunciation of West Usambara: Shambaa rather than Sambaa.

Usambara, especially its central plateau, had large tracts of dense forests, while the West, was more poorly forested (Krapf 1858: cited in Iversen 1991; Farler 1879; Eick 1896). Even the Mlinga block was stated to have much forest (Krapf 1858: cited in Iversen 1991; Meyer & Baumann 1888), although Farler (1879) also noticed many villages here. Johnston (1879), when staying on the Magila Hill at the southern foot of the Mlinga massif could observe forest-covered mountains closing the horizon from northeast to southwest. Bellville (1875) observed:

> Some of the trees were very lofty, over 100 feet and above six feet in diameter; the natives call them 'mvale[6]'. After ascending a good distance, the path left the gully, and passed through a small village, many of which are on the Mountains.

The Sigi Valley was claimed to have a dense, untouched rainforest, and on the road between Muheza and Derema the border between dense forests and grass and bushland ran at about 600m altitude (Scheffler 1901: cited in Rodgers 1993). However, as late as about 1880 it was claimed that dense forests reached all the way down to Muheza (Moreau 1935). A mapping expedition in 1888 gave reports of the area's vegetation (Meyer & Baumann 1888, Baumann 1889: cited in Iversen 1991). Dense forests were reported to dominate the whole plateau of East Usambara from Sigi to Lutindi, as well as in Mafi, Mlinga (except the steepest parts having grasslands), Segoma, Tongwe and Sigi Valley.

Conte (1996) identifies a precolonial ethic of conservation in Usambara, as well as techniques of resource preservation. He argues social relations among the Sambaa farmers and Mbugu herders permitted deforestation in some areas and conservation in others. It would appear that the forest was managed with respect to local custom (Table 2.2).

With the arrival of German colonists at the end of the nineteenth century the relatively stable pattern of land use changed, with people moving up the mountains to more inaccessible places (Rodgers 1993). German settlers carved out large estates in forested areas for coffee, sisal, rubber, oil palm and teak plantations (Meyer 1914: cited in Iversen 1991; Grant 1924; Forest Department 1930). Coffee plantations were first started up in the East Usambaras in 1891 (Milne 1937). Magrotto Estate on Magoroto Hill started in 1896 producing both coffee and rubber.

Schabel (1990) details the early history of German Forestry in

---

[6] 'Mvale' is probably ***mvule*** (*Milicia excelsa*).

Table 2.2 Evolution of approaches to natural forest management in the Eastern Arc Mountains

Period	State	Forest Interventions	Predominant Natural Forest Management Approach
1740-1892	Kilindi Kingdom and Hehe Chiefdoms	Conservation in some areas, deforestation in others	Local Customary
1892-1914	German administration	Reservation	Technocratic
1914-1919	First World War[7]		
1919-1961	British administration	Reservation	Technocratic
1961-1989	Independent	Reservation	Technocratic
1989-1998	Independent	Reservation, farm forestry	Participatory
1998-2000	Independent	Joint Forest Management and Community forest management	Political Negotiation

Source: Author's analysis of secondary data & see Table 1.1.

Tanzania. The first forester appointed in 1892 led to a full Department of Forestry and Wildlife by 1912 (Rodgers 1993). Volkens (1897: cited in Rodgers 1993) and Siebenlist (1914: cited in Rodgers 1993) stressed the need for forest reservation and even reforestation in German East Africa. Schabel (1990) summarised the main major German achievements. By 1914, 231 separate forest reserves were gazetted covering more than 7500 square kilometres. These were mainly mountain and coastal forest. German maps show the presence of settlements in some reserves, but Lundgren (1978) describing secondary forest in a few forest areas says that it is likely that most reserves were uninhabited. Whilst the Germans established many trial plots and aboreta of exotic plant species, for instance, Amani, they established very few plantations, totalling less than ten square kilometres. With the predominant implementation of forest reservation, the forest management approach could be identified as technocratic (Table 2.2).

The Germans evacuated Usambara in 1916 due to the First World War. With that evacuation, the forests became heavily degraded as locals started to clear the land for agriculture (Grant 1924). This was probably partly due

[7] With the First World War, political control and hence, forest management were uncertain.

to local people escaping enlistment and crop loss and the use of forest resources for emergency civil and military use (Rodgers 1993). However, Conte (1996) argues that the escalation in deforestation was a reaction to the German forestry policy of reservation. By clearing forest locals had learned to use European conceptions of land ownership as the basis for their own claims to land.

In the British administration, the Forestry Department of Tanganyika started in 1919 with the appointment of a Conservator to administer the reserves of the former German colony. In Usambara, some private German estates with much forest were converted to forest reserves (Parry 1962) and further areas of forest were reserved. In the 1930s and 1940s many areas of lowland forest were cleared for sisal estates. For instance, in the 1930s, approximately 800 hectares of forest in the Sigi Valley were cleared for the Sigi-Miembeni Sisal Estate. The predominant management approach could again be demonstrated as technocratic (Table 2.2).

With Independence in 1961, forest policies continued along the technocratic (Table 2.2) lines of the 1954 policy written under the British administration, with reservation of forest as the typical intervention (Wily 1997), although several forest reserves were degazetted. In 1962 it was proposed that Manga Forest Reserve should be degazetted or converted into a teak plantation. The proposal was never realised and in 1967 the Government of Tanzania still listed Manga as a forest reserve, although the eastern boundary of the reserve was moved back in order to give Mkwajuni villagers' 24 hectares for settlement purposes.

In the 1980s international interest in the Eastern Arc Mountain forests was renewed, with biodiversity increasing as a value of conservation. By 1989 interest was turned into action, specifically in East Usambara, with the implementation of two projects, the East Usambara Conservation and Agricultural Development Project (EUCADEP) and the East Usambara Catchment Forest Project (EUCFP). Initially interventions started off in the technocratic mode of reserving further areas of forest, but later turned to farm forestry and improved land use management, which are typical interventions of the participatory era (Table 2.2).

In 1998 the new national forest policy was announced, which empowers local community groups to manage forest. This policy therefore enables the development of forest management approaches in which roles can be politically negotiated (Table 2.2).

This brief history of natural forest management in Tanzania, and the Eastern Arc specifically, demonstrates the evolution of approaches and provides a framework for analysing stakeholders' changing roles via the predominant management approach of a particular period (Table 2.3). Chapter Four is organised via this framework, taking each management era in turn. It should be noted that although approaches may have been utilised simultaneously, it is the predominant approach that defines the era.

Table 2.3 Framework for analysis of findings via predominant natural forest management approaches

Predominant Natural Forest Management Approach/Era	Local customary	Technocratic	Participatory	Political negotiation
Period	1740-1892	1892-1989[8]	1989-1998	1998 onwards
Relationship to people	Forest management for the people and by customary leaders.	Management for the forest and against the people.	Forest management for and by the people.	Forest management with the people and other actors.
Interventions	Customary rules defining access and use of forest resources.	Forest Reservation.	Farm forestry, multiple resource use, buffer zones.	Joint Forest Management & Community-based Forest Management.

Source: Author's analysis of secondary data and see Table 1.1.

Stakeholders in the Eastern Arc Forests

This section preliminarily identifies and defines the stakeholders in Eastern Arc forests. The Tanzanian forest policy (The United Republic of Tanzania 1998) identifies seven main stakeholders as follows:

- Local communities;
- NGOs;
- Private sector and/or specialised executive agencies;

[8] For the sake of analysis, the end of the technocratic era has been identified as 1989, because it was at this time that EUCADEP and EUCFP started work in East Usambara and TFCG started their community-based projects.

- Local government;
- Forestry and Beekeeping authorities;
- Other government institutions; and
- International community.

For the sake of analysis, these seven have been amalgamated into five groups of stakeholders:

- Local communities;
- NGOs;
- State;
- Private Sector; and
- International community.

The State can be defined as an organised political community under one government. As a stakeholder it includes local government, the Forestry and Beekeeping authorities and other government institutions. The private sector includes estates and sawmills and the international community includes international researchers and donor funders. The local community and local NGO are defined more fully in the following.

Local Community

In most cases, the basic stakeholders in a forest are the people living within or adjacent to the forest, usually grouped under the term 'forest-local community' (or communities). Analysing who and what are the 'local community' with respect to Tanzania and the Eastern Arc Mountains specifically, is a prerequisite to analysing their specific roles in the management of natural forest.

The 'village' is perhaps the most ubiquitous social form of community in the world, although its institutional identity varies widely. In Tanzania, the village has served as the focus of socio-political development since the 1970s, where policies of grassroots self-reliance and co-operation (***Ujamaa***) were conjoined most definitively in the construct of the rural 'Village Settlement' (***kijiji***; plural, ***vijiji***), with particular features,[9] as summarised in the following (Wily 1997: 11-13):

---

[9] These features are embedded in law, originally in the *Villages and Ujamaa Village (Registration, Designation and Administration) Act*, No. 21 of 1975, superseded by a refined law of governance and administration, the *Local Government (District Authorities) Act*, No. 7 of 1982, introduced to formalise decentralised government (Wily 1997).

- The registered Village exists as a *discrete social community* with fully identifiable membership, for instance, it is not an open-ended society into which any person is able to randomly settle;
- The Village represents an integrated *socio-spatial unit*, 'Village' referring both to a definable social community and to the area of land they occupy and/or use;
- The Village represents a tangible *socio-institutional form*, and one that is recognised in law as a legal person able, for example, to sue and be sued in its corporate name; this corporate identity is held however not by the village community *per se* but by the government it elects to act on its behalf, the Village Council;
- These features combine to provide a framework within which principles of *common property* can be exercised in a statutory, not only customary, manner. The Village Council, as a definable agency and as a legal local entity, is able to own land and/or other resources, albeit in trust for its membership;
- The Village is *the foundation of national governance*; the Village Council is known in Kiswahili as *Serikali ya Kijiji,* or literally the Government of the Village; administrative law locates the Village Council as the most local level of Tanzania's formal and legally defined hierarchy of decentralised administration, albeit one which is subject to the direction of the next level of government, the District Council;
- The Village is an essentially *democratic and egalitarian institution*, operating by the governance of leaders who are elected by the entire adult membership of age eighteen years and above, not by those who attain authority through tradition, class or wealth, and not through appointments made by higher levels of government;
- The modern Tanzanian Village is a *viably sized working unit* of self-management that in turn enhances accountability in local management; a community cannot be registered as a Village unless it comprises 150 spatially cohesive households. Where it grows beyond a manageable range of around 300-400 households, it is legally assisted to sub-divide into two registered Villages, or to recognise Sub-Villages, each of which has a legally bound right to elect Sub-Village Chairpersons, who automatically sit on the wider Village Council.

Wily (1997) notes that the inability of the resource-poor central government to deliver basic services since the Villages Act of 1975 has served to consolidate local level self-reliance. In the post-colonial 1960s and 1970s, the majority of sub-Saharan African states dismantled local level or local customary organisation below the district level. Although highly varied in their success, Tanzanian villages of today exhibit a degree of organisational cohesion and productivity that is rare in Africa.

The 'Village Assembly' includes all member households, which are listed and established as the supreme authority of the community through registration. Member households from each sub village (***kitongoji***; plural, ***vitongoji***) elect sub-village chairpersons and all member households of the village, the Village Assembly, select the village chairperson and overall village government. These elected people are a representative government and form the 'Village Council' (***Serikali ya Kijiji***). The Village Council is issued with a certificate of incorporation.

Administrative law provides the Village Council with functional tools of management. The Village Council may form sub-committees, which may represent the Village in any government forum or court of law, for example, the Village Development Committee (VDC). The Village as a political unit via the Village Council is therefore able to make 'Village by-laws'. Once a Village by-law is drafted and approved by the local District Council (***Serikali ya Wilaya***), it becomes law, uphold-able in any court. The Village Council are granted 'Village Title Deeds', which means that they have 'Granted Right of Occupancy' granting the Village tenure for a period of 99 years.

Prior to Villagisation, farming communities tended to live in smaller settlements, much the same as the sub-villages of today. In fact, many of these settlements form the basis of sub-villages that have been aggregated to make the registered villages of today. Prior to the colonial administrations these settlements tended to be under the patronage of certain lineage or clan leaders and were locally centralised as chiefdoms (Conte 1996). In Usambara, which was politically centralised from around 1740, these chiefdoms became part of the Kilindi kingdom (Kimambo 1996). These local communities form the basis of the Village in Tanzania today.

Having analysed what is meant by 'local community' in Tanzania, throughout the volume, the word 'Village' or 'sub-village' shall be synonymous with 'local community' in respect to both the social community and the area of land belonging to the community. It is important to note that villages tend to be heterogeneous and in the analysis of stakeholders of any one forest the different stakeholders within that community and their separate interests must be examined. Since this volume analyses stakeholders' roles in the Eastern Arc forests in general there is no scope for a deeper analysis of community. Table 2.1 shows the forest-local communities for each forest case study.

Local NGO

The NGO movement is still relatively young and weak in Tanzania, as it had been customary that the single political party ran most sectoral activities from village to national level. Consequently, there was no niche or need for NGO activism or grassroots support organisations and although Tanzania had a long history of decentralisation, most natural resources were kept firmly entrenched in State control (Rodgers 1998). In the last two decades both local and international NGOs have increasingly played a role in natural resource management and conservation as it is increasingly recognised that the State has failed to manage natural resources effectively. The Tanzania Forest Conservation Group (TFCG) is one such local NGO whose activities have concentrated in the Eastern Arc Mountains and many of the case study communities in this research. The TFCG is identified as a stakeholder in the Eastern Arc forests, although perhaps in the majority of cases a secondary stakeholder rather than a primary stakeholder.

TFCG is a professional NGO rather than a grassroots NGO (Table 1.5). Using Cherrett et al's (1995) taxonomy of African 'environmental' NGOs (Table 1.5), TFCG has evolved from a conservationist to an environmentalist NGO. More recently TFCG is beginning to take on the role of facilitating grassroots, community-based NGOs and institutions to attain a meaningful role in the management of forest resources.

Chapter 3

Stakeholder Returns from Forest Resources

Forest Products

Whilst examining returns from forest resources the interface between forest (***msitu***), field (***shamba***) and bushland (***pori***) was often found to be blurred by local people. Hence, although this section is entitled 'forest products', both field and bushland derived products have also been identified by local people. In remaining faithful to the way local people identify returns from forest resources, the relationship between forest, field and bushland-derived products and local dependency on each of these resources for certain products can be examined.

Non-Timber Forest Products

Local community households in East Usambara utilise a variety of non-timber forest products (NTFPs). These have been identified as follows:

- Wild foods, such as: edible plants, edible fungi, fruit, honey and meat;
- Fibres for weaving mats and baskets, and for making brushes, thatch and rope;
- Medicinal plants; and
- Firewood.

Wild edible plants Local knowledge of wild edible plants, particularly green leafy vegetables, derived from forest (***msitu***), field (***shamba***) and bushland (***pori***) resources, is high in East Usambara. All groups and households interviewed and those observed through participation, regularly use wild edible plants in their diet. Twenty-three different varieties were named, seven derived from forest and 16 derived from field and bushland (Table 3.1) and in each case it is the leaves, which are eaten.

Table 3.1 Habitat of wild edible plants in East Usambara

Habitat	Family	Species	Vernacular Name
Forest	*Acanthaceae*	*Asystasia gangetica*	**tikini**
Forest	*Connaraceae*	*Rourea orientalis*	**kisogo**
Forest	*Menispermaceae*	*Dioscoreophyllum volkensii*	**msangani**
Forest	*Solanaceae*	*Solanum nigrum*	***mnavu***
Riparian forest	*Amaranthaceae*	*Alternanthera sessilis*	***mkoswe*/ mng'oswe**
Riparian forest	*Basellaceae*	*Basella alba*	***nderema*/ ndelema**
Riparian forest	*Convolvulaceae*	*Ipomoea aquatica*	***tarata***/talata
Field/bushland	*Amaranthaceae*	*Amaranthus spinosus*	***mchicha*/ bwache**
Field/bushland	*Capparaceae*	*Gynandropsis gynandra*	**mgangani**
Field/bushland	*Compositae*	*Bidens pilosa*	***mbwembwe*/ kisamanguo**
Field/bushland	*Compositae*	*Launea cornuta*	***mchunga*/ msunga**
Field/bushland	*Convolvulaceae*	*Ipomoea batatas*	***matembele*/ ukutu**
Field/bushland	*Cucurbitaceae*	*Cucurbita maxima*	**n'koko**
Field/bushland	*Euporbiaceae*	*Erythrococca kirkii*	mnyeumbeue
Field/bushland	*Euphorbiaceae*	*Manihot grahamii*	***kisamvu*/pea**
Field/bushland	*Lamiaceae*	*Platostoma africanum*	**kisungu**
Field/bushland	*Menispermaceae*	*Dioscoreophyllum volkensii*	**msangani**
Field/bushland	*Solanaceae*	*Nicandra physalodes*	**kibwabwa**
Field/bushland	*Solanaceae*	*Solanum gilo*	***nyanya chunvi*/ ngogwe**
Field/bushland	*Tiliaceae*	*Corchorus spp.*	***mlenda*/hombo/ kibwando**
Field/bushland	*Leguminosae*	*Vigna unguiculata*	**shafa**

Source: Author's fieldwork 1994-1998.[10]

All these species cannot be considered as truly wild since some of them are cultivated in home gardens and naturalised. Many species are introduced or pan-tropical. The former group is represented by some Solanaceae and Amaranthaceae species, which have originated in the

[10] Family and species names are taken from Vainio-Mattila et al (1997).

Neotropics. The latter group is represented by *Basella alba* (Basellaceae), *Bidens pilosa* (Asteraceae) and *Gynandropsis gynandra* (Capparaceae).

Collection of wild edible plants is solely the role of women. They are highly skilled in plant identification; for instance, **ndelema** (*Basella alba*) looks very similar to **matoyo**, which is a poisonous plant. Young girls learn from their mothers and female relatives at an early age which are edible and which are not. Collection is normally combined with other activities, for instance, on the return from cultivating or at the same time as collecting firewood. Only the best, young foliage is collected and usually only the amount required for that day.

Collection is seasonal in wealthier households who collect and eat wild leafy vegetables approximately twice a week. This may increase to three or four times per week in the dry season around January and February and between July and September when there is a reduction in the availability of planted vegetables. Households are required out of necessity to supplement their diet more frequently from the wild resource or buy extra supplies from village shops and Muheza market if cash is available in the household. However, the majority of households collect and eat wild leafy vegetables on a daily basis throughout the year. Daily dependency on wild leafy vegetables is often due to lack of access to alternative food sources through lack of money, distance to market and low ownership of poultry and goats.

The vegetable leaves often require further preparation on returning to the home or field where the vegetable is cooked as a side dish (***mboga***) to accompany ***ugali*** (a stiff maize porridge) or **bada** (a very stiff gooey cassava porridge). Women are very particular as to whether leaves of certain species are torn, cut or left whole and whether the petioles should remain or be discarded. Some plants, such as **mnavu** (*Solanum nigrum*) have thorns and therefore need careful removal before cooking. A number of varieties require a long cooking time, since they are bitter in taste and therefore may need between one to two hours cooking before they are palatable. **Msunga** is very bitter and therefore needs to be boiled and then squeezed of its bitter juices and then boiled again in fresh water. This procedure must be repeated three times before the leaves are palatable. Some vegetables are cooked in combination with others. The leaves are usually cooked with vegetable, groundnut oil or coconut milk and mixed with onion, tomato and salt. Some leaves impart a mucilaginous consistency to the sauce, for instance, **ndelema** (*Basella alba*) and **kibwando** (*Chorchorus spp.*).

There were differing opinions to preference for wild edible plants and those that are planted. Many expressed a preference for wild plants, saying that they are better in terms of taste, iron and protein content and abundance. A common statement was that the vegetables, "*increased the amount of blood in the body.*" Others commented, however, that planted vegetables saved on time of collection and food preparation.

Those communities that have access to public forest, such as Magoroto Hill and Kwatango, tend to prefer forest derived wild leafy vegetables (Table 3.2). In contrast, those adjacent to forest reserves and without public forest, such as Mkwajuni, tend to collect and eat field and bushland derived plants (Table 3.2). Through discussion with Magoroto communities it was discovered that forest derived plants are collected more frequently due to preference in taste. Those communities adjacent to forest reserve do not collect forest derived plants due to lack of access to the forest resource.

It was indicated that some of these edible plants are considered to be 'poor peoples' food. For instance, when talking to the women's group on Magoroto Hill there was much laughing and disgust at the mention of **tikini**. **Tikini** is available throughout the year and they say that they only use it out of necessity if other wild varieties and planted vegetables are not available.

Change through time in availability of different species was also noted. Older women mentioned on a number of occasions, that one variety of edible plant, known locally as **tebwa** (*Aerva lanata*), which was highly utilised by past generations is now very scarce and thought to be only found in some areas of the lowlands.

Few vegetables seem to have been domesticated around the home, with the exception of ***mchicha*** (*Amaranthus spinosus*), however since many are abundant in the fields and bushland, along paths and around the home this does not appear to have been necessary. It would seem however that those communities who do not have access to forest have adapted to utilise field and bushland derived edible plants.

Edible fungi The majority of households interviewed collect edible fungi. In East Usambara, nine varieties of fungi are collected and eaten from forest resources and 12 from field and bushland resources. Women usually

Table 3.2 Wild edible leafy vegetable preferences

Vernacular Name	Species	Habitat	Preference Rank[11]
Magoroto Hill			
nderema/ndelema	*Basella alba*	Riparian forest	1
msangani	*Dioscoreophyllum volkensii*	Forest	2
mchunga/msunga	*Launea cornuta*	Field/bushland	3
tikini	*Asystasia gangetica*	Forest	4
mbwembwe/kisamanguo	*Bidens pilosa*	Field/bushland	5
mnavu	*Solanum nigrum*	Forest	6
kisamvu/pea	*Manihot utilissima*	Field/bushland	7
mchicha/bwache	*Amaranthus spinosus*	Field/bushland	8
Mkwajuni			
***mchunga*/msunga**	*Launea cornuta*	Field/bushland	1
***mlenda*/hombo/kibwando**	*Corchorus spp.*	Field/bushland	2
***mchicha*/bwache**	*Amaranthus spinosus*	Field/bushland	3
***mbwembwe*/kisamanguo**	*Bidens pilosa*	Field/bushland	4
kisamvu/pea	*Manihot utilissima*	Field/bushland	5
kibwabwa	*Nicandra physalodes*	Field/bushland	6
Kwatango			
msangani	*Dioscoreophyllum volkensii*	Forest	1
***mchunga*/msunga**	*Launea cornuta*	Field/bushland	2
***mlenda*/hombo/kibwando**	*Corchorus spp.*	Field/bushland	3
***tarata*/talata**	*Ipomoea aquatica*	Riparian forest	4

Source: Author's fieldwork 1994-1998.

collect edible fungi, particularly those varieties that are collected from field and bushland resources. Those which are forest derived tend to be collected by both men and women when they are passing through the forest, but particularly by women whilst they are collecting firewood and edible plants. Collection is based on seasonal availability, therefore the majority of collection and utilisation occurs in the short (***vuli***) and long (***mwaka***) rains. Like edible leafy vegetables, fungi are eaten as an

[11] Preference ranking where 1 is the most preferred and 8 the least.

accompaniment to ***ugali*** or **bada** and are either boiled in water to make a soup or fried with tomatoes and onion. The most commonly collected edible fungi are those derived from field and bushland resources (Table 3.3) although those communities with access to public forest collect and eat a range of forest derived species.

Table 3.3 Most commonly collected edible fungi

Vernacular Name	Species	Habitat
nkuuvi	unknown	Field and bushland
vitundwi	*Termitomyces aurantiacus*	Cleared/Cultivated
magh'wede	*Auricularia delicatalpolytrichal fuscosuccinea*	Decaying Wood
ugenda na nyika	unknown	Field and bushland
magong'ongo	*Termitomyces letestui*	Cleared/Cultivated/ Edge of forest

Source: Author's fieldwork 1994-1998.

Wild fruits The forest is not a significant source of fruit for most members of the household. For most people it seems more convenient to obtain fruit from trees planted or retained in land clearance, in the fields and around the home, such as bananas, coconuts, oranges, mangoes, papaya, jackfruit, custard apple, grapefruit and guava. However, knowledge of forest and bushland derived fruits is extensive.

Twelve forest and bushland derived varieties of fruits were indicated as being utilised by school children (Table 3.4). It is the young school children who usually collect these fruits, eating half as they collect and returning with the rest for the family to share. Several parents stressed the importance of wild fruits in the diet of young children, since wild fruits and berries are rich in vitamin C as well as natural sugars.

Honey Honey collection is not traditional in East Usambara and is not widely collected. The majority of people know of the traditional bee keeping of Handeni Region, but do not know how to practice it. Many people are suspicious of bees and fear their stings. However, in each community there are usually a couple of people - all men - who are experienced in the collection of wild honey. One informant in Mkwajuni was experimenting with modern hives, selling the honey and wax to private individuals in Tanga.

Table 3.4 Wild fruits collected by school children

Vernacular Name	Species
maungo	*Cola sp.*
matoyo	*Unknown*
makonde	*Myrianthus holstii*
samaka	*Cola scheffleri*
matonga	*Strychnos nitis*
kwingwina	*Sorindeia madagascariensis*
vitoria/vitole	*Ancylobothrys petersiana*
makole/mkoe	*Grewia goetzeana*
vichaa	*Albizia schimperana*
maifae	*Unknown*
maviu	*Vangueria tomentosa*
matong'we	*Annona senegalensis*

Source: Author's fieldwork 1994-1998.

Those who collect from natural hives, collect mostly from tree hives and occasionally from underground hives. In the majority of cases no tree species were identified as being specific for hives, however Table 3.5 shows those that were identified by key informants. Hives are most often found in forest, but if found in the fields, the field owner will often ask the expert (***fundi***) to come and collect the honey. In return, the ***fundi*** will give the field owner a litre of honey and keep the rest for himself and his family.

Table 3.5 Tree species identified specifically for natural beehives

Vernacular Name	Botanical Name
***mbuyu*/muuyu**	*Adansonia digitata*
mkuyu	*Ficus capensis*
mvule	*Milicia excelsa*
muembe	*Mangifera indica*
mnyese	*Parkia filicoidea*

Source: Author's fieldwork 1994-1998.

In the dry season, between January and March, over five litres can be collected at one time, and at other time's perhaps only one litre. The beehives are often found inside the hollow of a tree trunk, and the honey is removed by hand after sedating the bees with smoke. A mixture of dry and damp coconut palms are said to produce the best smoke for this job. The collector is always careful to leave one or two combs remaining. In this way, the bees will remain and the collector may return at a later date.

Honey is normally collected for domestic consumption; however, if enough is collected then some may be sold locally. The bee larvae are often fried and eaten, but the wax is not utilised. Bee larvae contain ten times as much vitamin D as fish liver oil and twice as much vitamin A as egg yolk.

Wild meat Hunting of forest dwelling animals is an activity undertaken both in forest and on the ***shamba***. Hunting appears to fall into two main categories: in the ***shamba*** and in the forest. Protecting crops in the ***shamba*** by scaring and trapping animals is the role of the whole family; whereas hunting in the forest is solely a male activity.

Forest dwelling crop pests are a major problem for communities living adjacent to forest. A major investment of time is put into guarding the crops from these pests, particularly in ***vuli*** (short rains) and at harvest time. This is usually the role of women and children. Pests are scared away normally by the use of noise, throwing stones, catapults and using dogs, and occasionally using ***panga*** (bushknives), spears, and bow and arrows. Animals are rarely killed in this way, but if they are, they may be used for domestic consumption. The main attacks come from vervet monkeys (*Cercopithecus aethiops*) in the day and bush pigs (*Potamochoerus porcus*) and cane rats (*Thryonomys swynderianus*) at night. Birds, such as weavers, are a problem in the rice fields in the lowlands.

A number of farmers have managed traps on and around their ***shamba***. Many have experimented with placing snares and deadfall traps in and around the periphery of the ***shamba***. However they say that this is not effective since the animals are wise to this. Animals are rarely caught in this way, but of those caught, bush pig (*Potamochoerus porcus*) and mongoose (*Mungos mungo*) are most frequently caught like this.

Immigrants who live in Vumba, a sub-village of Kwatango, have an organised hunt once a week into the forest with the sole purpose of reducing populations of wild animals, which may attack their crops. They do not utilise the carcasses.

Villagers feel there has been an increase in crop pests in comparison to the past. Some suggest deforestation as the root cause, with less food causing the animals to search for food on the ***shamba***. Others have suggested that the increase is due to the fact that they are not allowed to hunt in the forest reserves, hence animal populations have increased.

Hunting in groups in the forest is solely a male activity. It occurs in all villages apart from in Mkwajuni village where no one admitted to this practice in Manga Forest Reserve, although on returning to the village at a later date two villagers were observed going into the Forest Reserve with guns. Each village has at least two groups of men who hunt together, once or twice a week and other groups who meet less regularly and see hunting as more of a social event. Groups of between three to ten men go into the forest with dogs, nets, bows and arrows, bushknives and occasionally guns. Very few people own guns, due to the expense of bullets. However, in almost every village there is one man who owns a gun. A hunter in Kwemnazi, sub-village of Mgambo, said that the younger generation seemed no longer interested in hunting in groups and feels that the last time of the great hunters was in the 1940s.

The most commonly hunted animals are bush pig (*Potamochoerus porcus*), blue monkey (*Cercopithecus mitis*), vervet monkey (*Cercopithecus aethiops*), red duiker (*Cephalophus natalensis*), bushbuck (*Tragelaphus scriptus*), and African civet (*Viverra civetta*). Other animals hunted include banded mongoose (*Mungos mungo*), rock dassie, colobus monkey (*Colobus angolensis*) and baboon (*Papio cynocephalus*).

Hunting in groups in the forest appears to be mainly a social activity. Wild meat tends to be for domestic consumption, and is divided equally among the members of the group and is only occasionally sold to the rest of the community. Sometimes the skins and horns are kept as trophies.

Trapping in public forest and forest reserves appears to be common, using both snares and pits. Traps have been noted in Manga Forest Reserve, although the villagers do not admit to trapping. Bush pig (*Potamochoerus porcus*), cane rat (*Thryonomys swynderianus*), banded mongoose *(Mungos mungo*) and common genet (*Genetta genetta*) are trapped most often. One individual in Kwatango manages fifty traps daily in one area of public forest. He normally traps approximately two animals per week, so returns appear to be low.

Communities do not depend on hunting as a source of protein or income. Rather than a necessity, wildmeat is either perceived as an occasional luxury and small income generator or by some - generally the younger generation - as 'old fashioned, poor peoples' food'. Whilst living in Kambai, I would eat wildmeat approximately six times a year. Cane rat was the easiest meat to buy with wildpig being less easily available. A leg of cane rat would cost approximately Tsh 200 (approximately 30 pence sterling) and would give a household of seven two full meals. Wildmeat is a cheaper option when compared to the price of a chicken, which in

Kambai would be sold at between Tsh 700 and Tsh 1000 (£1 and £1.50 sterling).

Hunting in groups is seen by some as an important male social activity. There are differences between tribes and religions in the type of meat that is acceptable to eat. Muslims obviously wouldn't eat pig. The Sambaa and Bondei tend to favour pig, cane rat and small antelope, whereas the Makonde are known to eat a wide variety of animals, varying from monkey and bushbaby to snakes and snails. The Sambaa and Bondei tend to refuse to eat baboon or monkey, since they feel they are too human like, and find the thought of snakes and snails revolting.

Hunters have noted that some animals are less abundant in the Magoroto Hill area than they were in the past. For example, the colobus monkey (*Colobus angolensis*) was last seen in the area in 1976. Most hunters attribute the reduction in animals in the area to their own activities and deforestation. They feel no remorse and feel that it is an advantage to have less wild animals in the forest in order to reduce occurrences of crop attacks.

Weaving materials The fronds of the wild date palm (*Phoenix reclinata*/***mkindu***/**msaa**) are used as a fibre (***ukindu***) for weaving baskets and mats for household use and wall hangings and food covers especially for wedding gifts. Individual leaflets of palm fronds are woven to make strips about two to three centimetres in width. These strips are sewn together side by side to make mats or coiled and sewn to make fans and wall hangings. Fans, food-covers, wall hangings, floor mats and baskets are the range of products produced. Some women gain extra income by making products to order locally.

Villagers say that ***ukindu*** is now difficult to obtain locally and ***mkindu*** (*Phoenix reclinata*) trees can only be found in the more inaccessible parts of sub-montane forest. ***Ukindu*** is collected locally on Kitulwe Hill and Mlinga Peak on Magoroto Hill, by an old man who goes hunting in those areas. The majority of material is bought from Tanga or Muheza markets, since it is whiter and considered to be of superior quality. Local ***ukindu*** is normally used for mats, which are to be dyed. The dye is also bought from Tanga or Muheza markets. Fans, food covers and wall hangings are often decorated with paintings of fruit and vegetables and a saying or proverb. In the villages food covers, fans, small and large mats can be sold for 250, 150, 1,000 and 5,000 TSh (30 pence, 20 pence, £1.20, £6.00 sterling) respectively and can be sold for almost double the price in the towns.

The fronds of the ***mnyaa*** tree are used locally for brushes and baskets and can be found in hilly areas, such as Kitulwe, Magoroto Hill. Bamboo, found on Magoroto and around Kambai is used to make larger baskets and crab baskets to be used domestically. Small baskets can be sold for 150 Tsh (20 pence sterling) and larger baskets sell at around 400 Tsh (50 pence sterling). The number of baskets made in a month depends on the amount of time available after working on the ***shamba***.

Thatch and rope The majority of houses are thatched with coconut (*Cocus nucifera*) palm fronds (***viungo***) and a small number of houses may be thatched with grasses and reeds (**nyasi: ufi** and **ngaghe, kinyusu, mainde**) all of which can be found around the ***shamba***. Wealthier families tend to use corrugated iron sheets, which extends the life of a house. Houses, which were built around the time of Ujamaa often, have corrugated iron roofs, since Nyerere offered people free roofing if they resettled in Ujamaa villages in Villagisation.

Almost all twine and rope for building comes from forest sources in the form of climbers, for instance, **usisi** (*Tiliacora funifera* and *Triclisia sacleuxii*) or tree bark, for instance, **ugoroto** (*Landolphia kirkii*). Sisal fibre, locally known as ***ukonge*** or **kamba** (*Agava sisalana*) is used only on rare occasions but is not considered durable.

Medicinal plants In terms of healing illness, local people tend to try a series of healing medicines (***dawa***), starting from local medicine, to sorcery and finally to western medicine.

Initially, local medicine (***dawa ya kiyenyeji***) will be tried. The patient may be treated by him/herself or their friends or family using plants known and found locally around the home and on the ***shamba*** and in bushland. The majority of people know of at least two or three plants, which can be used to treat minor ailments at home. Over 50 percent of households on Magoroto Hill actively retain plants around the home, which are useful medicinally (Table 3.6). It is useful to note the ailments which are most commonly treated, namely gastro-intestinal problems, chest problems and children's illnesses.

If the ailment is not eased by home treatment then the services of a ***mganga*** (local healer) will be required. The majority of families know at least one ***mganga*** and each village has at least three. Local medicinal plant knowledge is usually passed from father to son, however there are women who practice as ***waganga***, the majority of which specialise as midwives.

Table 3.6 Plants retained around the home for medicinal purposes

Vernacular Name	Species	Ailment
mzugwa	Unknown	Malaria/Intestinal Worms/Toothache
muuka	*Microglossa densiflora*	Intestinal Worms/Earache
hozandogoi	*Hyptis pectinate*	Intestinal Worms
vumbapuku	Unknown	Chest Cold/ Stomach Ache
mwarobaini	*Azadirachta indica*	Intestinal Worms
mvuje	Unknown	Prevent Cold
kivumbasa	Unknown	Stomach Ache/Toothache
mswele/mswelo	Unknown	Epilepsy in children
eeza	Unknown	Stomach Ache

Source: Author's fieldwork 1994-1998.

Waganga interviewed, suggested that the younger generation are increasingly less interested in learning local healing. Yet others move away to work in the towns and cities.

People of all faiths visit the ***waganga*** and the ***waganga*** say that they are as busy as ever. Normally, one ***mganga*** will treat between ten to twenty patients a day. Patients come from as far away as Mombassa to visit well-respected doctors, such as Benjamin Maua from Bombo, sub-village of Mgambo. Payment for treatment depends on the ability to pay. As Benjamin Maua said: "*If a patient has ten shillings* (eight pence sterling) *that will be enough.*"

Trees, shrubs, herbs and climbers are utilised (Appendices 1 & 2). The roots and leaves are used more often, but bark, seeds and fruit are also used. Most specimens are collected from ***shamba*** and bushland resources, but some can only be found in forest habitats (Appendices 1 & 2). Some ***waganga*** collect wild seeds and plant useful species around the home. Forest degradation caused by medicinal plant collection is minimal, since the majority of specimens are collected from bushland and ***shamba***. Even when specimens are collected from the forest, only small quantities of plant parts are required per treatment and plants are not killed in the process.

When staying in Mgambo, regular visits were made to Mzee Benjamin Maua who was a ***mganga*** and lived in the sub-village of Bombo. Each day there would be a group of approximately ten people waiting for him to treat them. On one occasion there was a patient who had come all the way from Mombasa to be treated by him. Usually, Maua would treat his patients inside his house, but occasionally he would treat them at his cave, where he took me and a patient once. The cave was approximately five minutes walk from his house, in his ***shamba.*** Once inside the cave, Maua wrapped a black piece of cloth around his body - the uniform of a ***mganga*** - and began to sing a song in Kisambaa, whilst tapping the root of the ***mjolwe*** tree with a knife. He asked the patient to lie on a mat on the floor while he sang and prepared the medicine. From inside two large antelope horns, he took powder which was wrapped in small pieces of cloth and mixed them together and wrapped it in a piece of paper. He then told the patient to take a pinch of the mixed powder in his ***uji*** (maize porridge); three times a day for three days, to prevent epileptic fits over the following six months.

If the illness is not treated successfully through medicinal plants used for healing, the patient may then seek the assistance of a ***mchawi*** (witch) who specialises in the use of medicinal plants for sorcery or witchcraft. Sorcery or witchcraft is used to treat illness when it is thought that illness itself has been caused by a ***mchawi.*** The patient will require the assistance of a ***mchawi*** whose ***dawa*** is thought to be stronger than the ***mchawi*** who originally applied the curse. In these cases, it is usual for the fee to be a chicken or a large sum of money.

The majority of households invest in protective medicines for a range of problems. At the door to many homes, a small bottle of protective medicines is buried to ward of evil spirits (***mashetani***), and disempower ***machawi.*** Husbands, who work away from home, often bury ***dawa*** at the entrance to the home, to prevent other males from entering the house to have sexual relations with their wives. Similarly, ***dawa*** is often placed on pathways leading to the ***shamba*** to protect against curses of the land. Small children often wear necklaces, bracelets or anklets, of string or leather with shells, horns, seeds, or small pouches of cloth, which contain ***dawa*** to ward off harmful ***mashetani*** or ***machawi***.

In Kambai there are a number of individuals that are famous for their 'magical powers'. Msukuma was one man who was known to have powers over a very large black mamba. If Msukuma was away from his ***shamba***, it was thought the snake would protect the ***shamba*** from thieves or curses. Another old man held special ***dawa*** that enabled him to fish or cross-rivers when crocodiles were present. He is said to place the ***dawa*** in his mouth as he enters the water. He would then call the crocodiles that would come to him and leave the water, so that he could cross or fish safely. A number of elders would borrow this ***dawa*** when they wanted to fish in a part of the river, which held crocodiles. There were a number of old men who were genuinely feared if crossed, one of which was Mzee Ngobegobe. This man was said to have been exiled from West Usambara for using witchcraft to murder whole families. He was said to be able to change his appearance at will, to that of an owl. When he died, most people thought that it was because he had tried to use his powers against someone who had strong protective medicines. His ***dawa*** was thought to have turned back on him and it was thought that that was what had killed him.

Witchcraft is part of everyday life in Tanzania and there is ***dawa*** for ultimately anything and everything: from murder to love. It is also held up in court. A friend Mary, is the third wife of a very powerful witch. He was taken to court and fined for attempted murder through witchcraft. He believed his second wife was having an affair with a young man in the village. He and a friend ambushed the young man and forced him to eat an oyster nut along with ***dawa*** mixed with his own blood and that of the wife's. The young man took Mary's husband to court for attempted murder, since it was believed that if he had been having an affair he would have died upon eating the oyster nut. The Makonde women are also believed to have powerful magic, which enables them to control their husbands drinking and money spending better than the Sambaa. I was also invited to watch a ***mchawi*** complete a ceremony where he gave young men the ability to become invisible if they wanted. The young men, wished to emigrate to Germany, but feared immigration officials.

If neither a ***mganga*** or a ***mchawi*** appear to cure an illness, it is then that villagers go to the hospitals or dispensaries for western medicine (***dawa ya kisasa***). Magoroto Hill has one medical dispensary, situated in Mwembeni village. For more serious ailments, villagers travel to Kicheba or TEULE hospital, Muheza. When Magrotto Estate was in production, ill people were transported by the estate vehicle to TEULE. Ill people are now carried down the hill by stretcher. Mkwajuni and Kwatango villagers use Lanzoni and Kwatango dispensaries respectively. Kwatango dispensary is supplied with a monthly package of drugs donated by UNICEF, but villagers say that supplies do not last the month. Kambai has no dispensary, but has small shops (***duka***) that sell small amounts of malaria drugs. Otherwise, villagers would have to travel to Kwatango or Kiwanda to the dispensary, where they would often return without help since there were no drugs remaining, or walk the five hours to Muheza TEULE hospital.

This series of healing options was the norm: local healing, witchcraft and then western medicine. However, most ***mganga*** and local people recognised that there were some illnesses that ***dawa ya kiyenyeji*** could not assist, namely malaria and patients would either go directly to hospital or be advised to use western medicine. There are also occasions when western medicine and local healing are used side by side in combination, for instance, local midwives working together with Western medically trained midwives. The power and fear of witchcraft is very strong however and there have been two occasions where prolonged pain was witnessed. One was when a friend Grace had her third consecutive miscarriage. There were complications and she was taken to Muheza hospital in the Landrover. Unfortunately she was delayed for three hours, while the ***mganga*** worked to remove a curse that they believed she had. When the ***mganga*** failed she went to hospital where she was assisted. On another occasion, a neighbour Christopher who was one of the wealthiest people in the village came to the house crying and asking if he could be taken to town urgently. He had a small boil on his foot, that was painful and he believed that a jealous person had put a spell on the path from his ***shamba***, which had caused him to get the boil. He wanted to be rushed to town, but would not go to the hospital, because he wanted to see his ***mganga***, who was very powerful. He was taken to town and dropped off at the house of the ***mganga***. The next day he was seen at the hospital, where he had been treated. He gave an embarrassed grin and said that he was feeling much better.

Firewood Fuelwood is by far the most commonly used timber forest product in all locations, whether frequency or volume measures consumption. The main users are domestic but consumption by small restaurants (***hoteli***) is also important.

In the areas studied, firewood is the main source of fuel. The three-stone fire is the cooking system used by all. Kerosene is used by only a small number of wealthier families, who may have a small stove, but often it is only used to help ignite the fire. Charcoal is not utilised due to the lack of locally experienced producers and the availability of firewood.

Firewood is collected exclusively by women. Men may bring a piece of wood back to the home on occasion when they are passing through the forest. Collection patterns are determined by family size, number of women in the household, proximity to the source of firewood and its ease of collection. If good quality firewood is readily available close by to the home or ***shamba***, then women tend to make frequent trips, collecting fairly small amounts each time. If firewood is scarce, trips are less frequent, but more women are involved in one trip and very large bundles are carried. On average women collect two to three times per week. Firewood collection is often undertaken on the return from the ***shamba*** and the workload is divided up between the female members of the household.

The majority of firewood is collected from forest resources, however communities without access to forest resources, more usually collect from their ***shamba*** and bushland, although they do admit to occasionally stealing from reserves. The reason given for stealing from reserves was that the next nearest source is one to two hours walk away.

The time taken to reach the forest from which firewood is collected ranges from ten minutes in the case of Mgambo to two hours in Mkwajuni. Villagers say that it is becoming increasingly difficult to find good quality firewood and they are now required to travel deeper into the forest. On arrival at the forest, the distance walked in from the edge ranges between five minutes and forty-five minutes. By the time firewood has been located, bundled up and carried home, the total journey time can range between fifty minutes and four hours.

In most places only dead wood is taken for firewood. Women in Mgambo were found to be collecting wood from old pit-sawing sites. Large logs are split into more easily managed pieces. All women are highly selective about the species they prefer to collect (Table 3.7), with only recent immigrants to the area having difficulty in naming the species they prefer. If there are abundant supplies then women can afford to be selective with the species taken, and are well aware of the relative calorific values of the different types of wood available. Other characteristics noted as being desirable for firewood are ease of ignition, small production of smoke, taste and smell of smoke, slow burning, splitability and ease of rekindling the fire.

In Mabajani, sub-village of Mwembeni, where there is a perceived shortage of firewood, women said that they use crop residues such as coconut husks and maize cobs, they also try to cook only foods that have a relatively short cooking time. Crop residues were also utilised in Bombo, sub-village of Mgambo, when the women have no time to go collecting, particularly when they are ill. It was found that no households buy or sell firewood, there being no market.

Each community has at least two ***hoteli*** where tea and doughnuts (***chai na mandazi***) are prepared. Larger ***hoteli*** may serve a wider range of food, such as bread, ***ugali*** and beans. An ***hoteli*** is normally attached to the owner's home, but the kitchen area is separate from the families. One three-stone fire is used and kept burning most of the day to keep water hot. Firewood is usually collected by the owner, with small children sometimes assisting.

Table 3.7 Tree species preferred for firewood

Vernacular Name	Botanical Name	Preference Ranking[12]
Magoroto Hill		
mgwiza	*Bridelia micrantha*	1
mshai	*Albizia adianthifolia*	2
mohoyo	*Afrosersalisia cerasifera*	3
msambia	*Pachystela msolo/brevipes*	4
***msaji*/mjohoro**	*Cassia siamea*	5
mkongoo	*Sapium ellipticum*	6
Mkwajuni		
mkole/mkoe	*Grewia goetzeana*	1
***mbwewe*/mftumba**	*Lecaniodiscus fraxinifolius*	2
***msewezi*/mseezi**	*Faurea saligna*	3
***mtalawanda*/mtaanda**	*Markhamia hildebrandtii*	4
msagusa	Unknown	5
Kwatango		
mhande	*Craibia zimmermannii*	1
mfumba	*Carissa edulis*	2
***mbwewe*/mfumba**	*Lecaniodiscus fraxinifolius*	3
***mtalawanda*/mtaanda**	*Markhamia hildebrandtii*	4
***mwawia*/muawia**	*Xylopia sp.*	5

Source: Author's fieldwork 1994-1998.

Timber Forest Products

Local community households in East Usambara utilise a variety of timber forest products (TFPs). These include:

- Building poles; and
- Timber for making furniture, and household and agricultural utensils.

Building poles All villagers construct their own houses, the only exceptions being estate workers and catchment forestry guards who often have accommodation provided with their posts and perhaps some school

[12] Preference ranking where 1 is most preferred and 6 is least preferred.

teachers who do not originate from the area and may rent accommodation. Rented accommodation costs approximately 400 Tsh (50 pence sterling) per month for a two to three bedroom house with kitchen and main living room.

Required building materials include timber, rope, plasterwork, bricks and roofing thatch or corrugated iron sheets. The standard home is a two-bedroom house, with a living area and a separate building for the kitchen and toilet. The majority of homes are built from poles with mud daubing and grass or coconut palm thatch. The wealthier households may have corrugated iron roofs; smooth, whitewashed plasterwork or sun dried or fire burnt bricks all that extend the life of a house. Repairs to plasterwork are needed a couple of times a year and some pole replacement may be required after four or five years.

The local building style requires standard poles or timbers cut roughly into long thin pieces to do the same job. Building poles are exclusively collected by men. Live trees are most commonly utilised for building construction, replacement and repair. Poles are cut from saplings for the withies, which are approximately two centimetres in diameter and two and half to three metres in length. Larger trees are often taken and split for the vertical poles and beams, whose diameters range between ten to 15 centimetres in diameter and two and a half to three metres in length. Table 3.8 shows the average building pole requirements for typical village buildings.

Table 3.8 Average building pole requirements

Type of Building	Size of Room (metres)	Vertical	Number of Poles Horizontal	Beams
Two room house	3.5 by 3.5	70	200	60
Toilet	1.5 by 1	20	70	25
Kitchen	3 by 3	40	100	30

Source: Author's fieldwork 1994-1998.

Most people are well aware of which tree species make the best building poles (Table 3.9). Characteristics include rooting ability, termite resistance, durability, hardness and straightness.

Table 3.9 Tree species preferred for building poles

Vernacular Name	Species	Preference Ranking[13]
Magoroto Hill		
Mgwiza	Bridelia micrantha	1
Msaji/mjohoro	Cassia siamea	2
Msambia	Pachystela msolo/brevipes	3
Mshai	Albizia adianthifolia	4
Mpera/mpea	Psidium guajava	5
Mnyasa	Newtonia buchananii	6
Mkwajuni		
Mtalawanda/mtaanda	Markhamia hildebrandtii	1
Mkole/mkoe	Grewia goetzeana	2
Mbwewe/mfumba	Lecaniodiscus fraxinifolius	3
Mlanga	Millettia sacleuxii	4
Msewezi/mseezi	Faurea saligna	5
Kwatango		
Mtalawanda/mtaanda	Markhamia hildebrandtii	1
Mkeakiindi	Diospyros mespiliformis	2
Mbwewe/mfumba	Lecaniodiscus fraxinifolius	3
Mhafa	Millettia dura	4
Mnkande	Stereospeirmum kunthianum	5
Mkuyu	Ficus capensis	6

Source: Author's fieldwork 1994-1998.

Mkwajuni villagers expressed favour for *Brachylaena hutchensii* (**mkarambati**) which was once abundant in the area but is now unavailable due to overuse. People say they are finding it increasingly difficult to obtain good quality building materials from public forest and some say that they have to steal it from the forest reserves.

Wood for furniture and utensils Pitsawing in the forests has been banned since January 1993. However, a number of pitsawing sites were seen on the south-eastern boundary of Magrotto Estate, which were reported to have been undertaken in April 1994. In Kwatango and Kambai problems with

[13] Preference ranking where 1 is most preferred and 6 is least preferred.

pitsawing were occurring up until 1997, with the Kambai government forest reserve guard illegally pitsawing in Kambai public forest.

The majority of pitsawyers tend to be under contract from businessmen in Tanga, who then export the timber to Kenya and Dar es Salaam. In general pitsawyers are not local and live and work within the forest and provide little extra income for locals, unless a tree is on a ***shamba*** and then the owner will be paid a proportion or more often given a plank of timber.

The best timber species are ***mvule*** (*Millicia excelsa*), ***mbambakofi*** (*Afzelia quanzensis*) and **mnyasa** (*Newtonia buchananii*). A Miembeni villager who had previously been a pitsawer told me that although there was a ban on pitsawing due to over exploitation of these species, the District Forest Officer had given permission to harvest **mnyasa** (*Newtonia buchananii*) for school desks for the four schools of Magoroto Hill. Table 3.10 shows prices (1992) received from timber of one tree when sold to the traders from Tanga.

Table 3.10 Monetary returns from pitsawn timber (1992)

Vernacular Name	Species	Cash return from timber of one tree
mvule	*Milicia excelsa*	300,000 TSh
mbambakofi	*Afzelia quanzensis*	180,000 TSh
mnyasa	*Newtonia buchananii*	150,000 TSh

Source: Author's fieldwork 1994-1998.

Most furniture is purchased from local carpenters and is made to order from pitsawn timber. Carpenters expressed concern over the pitsawing ban and said that they would be forced to stop production when existing stores were depleted. Many households own planks of timber, which they store in the rafters of their homes for furniture required in the future.

Mvule (*Milicia excelsa*) and ***mbambakofi*** (*Afzelia quanzensis*) are the preferred tree species for furniture making because they are hard and resistant to termite attack. **Mnyasa** (*Newtonia buchananii*) is also commonly used because it is easily available on Magoroto Hill. It is easy to work with and is used for windows and doors, hoes and machete handles, but is not suitable for furniture.

Beds, chairs, doors and windows are the most common furniture produced. The market is predominantly local, however there is a small market for selling to locals who are now living in the towns (Muheza,

Tanga, and Dar es Salaam). These people prefer to have things locally made, because it is cheaper than in the towns. A part-time carpenter can make a profit of approximately 89,250 TSh (approximately £110 sterling) per annum (1993).

Other wooden household utensils required include stools, ***mbuzi*** (used for grating coconuts), large pestle and mortars for grinding maize and dehusking rice, and spoons and ladles, along with agricultural tool handles. Men within the household usually make these implements.

Other articles made from wood include drums and other musical instruments and small wooden bicycles, all of which are usually made by young men and boys who use them.

Forest Services

Returns from forest resources include not only the forest products, but also the services gained from forest resources, 'forest services'. Forest services include water catchment and soil conservation, ecological services that the government has given great importance to through the East Usambara Catchment Forest Project (EUCFP) and which local communities understand well. Yet, forest services also include forest land, the land on which forest grows. Forest land is highly valued when it is converted to agricultural land.

Conversion of Forest Land to Agricultural Land

The known settlement history of Kambai is relatively recent when compared to the mountain settlements of East Usambara. However, as discussed in Chapter Four, it is possible that the name Kambai means in Kisambaa 'the home of the Kamba people'. The Kamba were elephant hunters and ivory traders and often had trading posts in the lowlands where they could exchange meat and ivory with the Sambaa of the mountains. Since the Zigua had driven the Kamba out of the Luengera valley in the early 1800s, perhaps some relocated in the Sigi-Valley at Kambai where they could trade with Sambaa of the Amani block.

Local history tells of the Sambaa of the Amani block resettling in the forests of Kwezitu (literally 'to the dense forest') and Kambai in order to escape German rule and enlistment in the First World War. It is thought locally that many of those Sambaa escaping enlistment were guided to the Kambai area by the sound of Zumbakuu waterfall, whose sound as the water fell into an underground cave was thought to have been heard 40 kilometres away in Tanga. Some informants believe that perhaps these were the first to settle and clear small areas of forest for farm land.

In the 1930s the British cleared 800 hectares of forest for the Sigi-Miembeni Sisal Estate, where they operated Mgambo Sawmill also. The estate brought in immigrant estate workers and it is the individual histories (Boxes 3.1, 3.2, 3.3, 3.4 & 3.5) of these immigrants which demonstrate how the forests of the Sigi Valley were settled, cleared and farmed and the present day settlement and farming pattern of Kambai was created.

Box 3.1 Oral history of Mzee Daili Maneno of Kambai

"I am Makonde by tribe and was a ***manamba*** (casual labourer). The rich men came to our region to convince us to go to Tanga Region to work on the sisal estates. We were cheated. I first worked in Korogwe at Kwalukonge Sisal Estate from 1954. I worked there for only two years and found that the work was very hard. So I moved to Pangani to work on another sisal estate, where I stayed for only one year. Then I heard that my brother-in law was here at Sigi-Miembeni, so I came here in 1956 or 57. Later I left the work and wanted to cultivate.

The Kambai villagers who then resided around where the school is now, showed me a forested area which was under Mgambo Saw Mills, but they were not working in that area. I cleared a ***shamba*** and began to cultivate. In 1964 the manager from Hale Sisal Estate told villagers that they must stop cultivating the land, because he had bought it and wanted to plant sisal. The villagers very quickly built a school and a strong village. So we ran to the government and the government said we could stay."

Source: Author's fieldwork 1994-1998.

These people were the initiators of Kambai as an Ujamaa village. The majority settled in Kambai itself, but others stayed at Msakazi (literally 'for work') sub-village where they utilised the estate buildings and others settled at Kweboha (literally 'for alcohol') sub-village. Both Msakazi and Kweboha are on what is now SHUWIMU land, so all settlers are theoretically squatters (Box 3.2). Many of Kweboha's population are Sambaa who originated from West Usambara and moved to Kwezitu and Msasa IBC villages on the Amani block for tea estate work and later moved down to Kweboha to farm (Box 3.2). The majority of people settled at Kambai, however, where each male household head was given one acre (just over half a hectare) of public forest on which to farm (Box 3.4).

Box 3.2 Oral history of Mzee Charles Mhina of Kweboha, a sub-village of Kambai

"I am ***Sambaa*** and came to the Kambai area in 1950 to work at the Sigi-Miembeni Estate. I was an accountant. I stayed for five years and worked under two managers from Europe - the first being John Powers and the other Mr. Sawmill! Then an Asian manager came and he made a lot of 'noise'. I couldn't work well with him. So after five years I left to work as an accountant at the Amani tea estate, Bombay Bwima until 1972 when I retired.

I built my houses and farmed in Kisiwani in Amani. I planted cardamom and coconuts and lived very well. Then in 1977 I was chosen to be village chairman. The villagers chose me to collect money for buying and selling cardamom. One day I went to Muheza to collect 300,000 TSh, but on the way some people had planned to ambush me and steal the money. They failed, but I decided it was too dangerous to carry on with this work. I made an excuse to the villagers that I was too ill to carry on with the work because I had high blood pressure.

Some of the villagers became jealous of me in Kisiwani, because I was a successful farmer. A lot of ***machawi*** (witches) tried to use spells against me. So in 1987 I moved to Kweboha. I had already worked in the area, so knew the land was fertile.

Although I am living here, I'm afraid because I'm on SHUWIMU land. However, I have seen the majority of us, we are planting major crops. Also the DC (District Commissioner) has told us not to worry about planting trees: 'Plant as many as you can.' The DC said that."

Source: Author's fieldwork 1994-1998.

Those who arrived later had to clear remaining forest on what had become SHUWIMU land (Box 3.5).

Whilst the estates were still in production food was required for many of the workers who had little land and little time to produce enough crops for themselves and their families subsistence. This again brought people into the area that cleared and cultivated land for cash crops to sell to the estate for food. Informants on Magoroto Hill also describe the clearing of forests in the 1940s: "*the land was cleared and cultivated right up to the hilltop. Successful farmers were producing excess food for sale to immigrant estate workers*."

Box 3.3 Oral history of Benjamin Nantipi of Kambai

"I am a Makonde and I was born near Lindi. In 1930 my parents and I moved to Tanga Region to work on the sisal estates. We first came to Msakazi (sub-village of Kambai) to work on the Sigi-Miembeni Sisal Estate. We later moved to Dar es Salaam and Morogoro. Eventually we became tired of moving around and returned in the 1950s to Msakazi. I became chairman in Msakazi in 1954. I left in 1962 to move to Kambai to start farming. Later I became chairman of Kambai Village."

Source: Author's fieldwork 1994-1998.

Box 3.4 Oral history of Linus Erenso of Kambai

"I originate from Mbeya and first moved to Morogoro to work on a sisal estate there. Later I moved to work on the Sigi-Miembeni Sisal Estate. In 1969 I moved to Kambai to cultivate and to start the Ujamaa village. When starting the Ujamaa village everyone was given one acre of forest land on which to farm. I had lots of goats, some neighbours' thought that my goats would eat their crops. I persuaded them to sell their plots, so that I could have four acres next to my home!"

Source: Author's fieldwork 1994-1998.

Box 3.5 Oral history of John Mponda of Kambai

"I left Iringa when I was seventeen years old to work on the tea estates in Amani. I worked there for two years, but saw that the work was very difficult and hard. I had heard about Kambai and so moved there in 1974. I found it difficult to obtain any land, so one elder advised me to clear some forest from SHUWIMU land. So in 1975 I cleared seven acres for myself."

Source: Author's fieldwork 1994-1998.

Since the 1980s immigrants to Kambai have been advised by the elders to clear village land which extends from the village to the Muse River (Boxes 3.6 & 3.7).

Kambai and Semdoe Forest Reserves were fully gazetted in 1993 and 1994 respectively. There have also been rumours that SHUWIMU want to start an orange estate (Nipashe 1996) and that they would ask Kweboha

Box 3.6 Oral history of Agnes Christopher of Kambai

"We started to farm the land in Kambai in 1986. Christopher had a small farm, which he extended into the forest and bush surrounding it. We now have six hectares. We decided to move to Kambai, because Christopher is a timber merchant and had been to the area to pitsaw in the past and knew the area to be fertile."

Source: Author's fieldwork 1994-1998.

Box 3.7 Oral history of Msukuma of Kambai

"I come from Shinyanga and am of the Msukuma tribe. When I was young my mother married another man and moved with my sister to the Sigi-Miembeni Estate in 1951. When I was older I moved to Tabora where I had been told my mother was. I stayed and farmed for a very long time.

Then one day my sister asked my mother if they could get money for bus fares to Shinyanga to see if their family was alive or dead. They went but could not find anyone. Then someone told her that I had moved to Tabora. They found me and persuaded me to get the bus fare to Tanga and leave everything in Tabora because they had a very good farm in Kambai.

On arriving in Kambai I found that they had a very small farm and were very poor. I was unable to turn back and without anything I had to work on other peoples' farms to earn money. In 1991 the elders advised me to clear forest and bushland towards Muse. I now have four hectares."

Source: Author's fieldwork 1994-1998.

and Msakazi residents to relocate. The present day community of Kambai is an 'island' landlocked by Kambai and Semdoe Forest Reserves and SHUWIMU land. Benjamin Nantipi, Kambai Village Chairman said in 1996: "*Villagers need land. Without SHUWIMU land, there is no longer anywhere to go but the forest.*" In 1997 the Village Government attempted to strengthen the community's claim on SHUWIMU land by supporting the clearing of SHUWIMU land for agriculture by 15 Kwezitu villagers who requested permission to take one hectare each of land for maize cultivation. The Village Government were unable to give permission, but gave their support and were themselves supported by the government at Ward and District level.

Returns from Forest Land Converted to Agricultural Land

In the conversion of forest land to agricultural land, forest is cleared and the bushes and grasses are slashed and burned. Not all the trees are cleared, since it is customary to retain some species in fields. Table 3.11 shows those tree species most commonly retained in the fields on clearance of forest. Some tree species are retained because they are protected either by the government (National trees) - locally known as 'government trees' - for example, ***mvule*** (*Millicia excelsa*) or through custom, for example **mkuyu** (*Ficus sycomorus*). Others species are retained for their returns, for instance **mtalawanda** (*Markhamia hildebrandtii*) and others are simply retained because they are too big or hard to cut easily.

Villagers not only retain trees when clearing forest land; they also plant trees, both in their fields and around their homes. Prior to TFCG intervention in Kambai village, villagers had planted only exotic tree species (Table 3.12). These were either from wildlings that had become established in forest and the fields of friends and family or seedlings that had been given from friends and family who lived elsewhere and had obtained seedlings from EUCFP or EUCADEP.

In 1994 TFCG started to work in Kambai village, predominantly with the intervention of farm forestry. The tree species to be raised on the TFCG nursery were identified by villagers through both individual and group SSIs and matrix scoring and ranking. Table 3.13 shows the tree species planted after TFCG intervention. By 1997, 62 Kambai villagers had bought and planted 96,322 tree seedlings from TFCG. Of the total seedlings bought and planted, 82 percent (79,088) of tree seedlings survived. These figures are only a fraction of the total number of seedlings planted in these three years, since TFCG assisted individuals and groups to manage their own tree seedling nurseries. Several individuals and groups raised enough seedlings to sell locally also. These individuals and groups were able to sell their seedlings at higher prices than TFCGs subsidised prices, because after the first year, TFCG reserved their seedlings for first time buyers of seedlings and those who had little cash and obtained seedlings by exchanging seeds collected from the forest for seedlings. Those that had bought seedlings previously or were relatively well off were expected to produce their own seedlings in nurseries or buy locally from other villagers.

Villagers have planted trees in their fields either as woodlots, on field boundaries, intercropped with maize or as individual trees in the field or around the home. Three different categories of reasons for tree planting

Table 3.11 Tree species commonly retained in fields

Vernacular Name	Tree Species	Reason for Retaining in ***Shamba***
mvule	*Millicia excelsa*	Timber & shade (Government reserved tree).
mgude	*Sterculia appendiculata*	Plywood (Government reserved tree).
mkuyu	*Ficus sycomorus*	Soil conservation, soil improvement, & water retention (Customary reserved tree).
mvumo	*Ficus thonningii*	Soil conservation, soil improvement, water retention (Customary reserved tree).
mshai	*Albizia gummifera*	Timber, shade, soil conservation, nitrogen fixation.
mtalawanda	*Markhamia hildebrandtii*	Building poles, firewood.
mgolimazi	*Trichilia emetica*	Timber, firewood, shade, soil conservation.
mbuyu	*Adansonia digitata*	(Customary reserved tree).
msufi mwitu	*Bombax rhodognaphalon var. tomentosa*	Plywood, stuffing for pillows and mattresses.
mtondoro	*Julbernardia globiflora*	Timber for making tukulanga drums.
mtambaa	Unknown	Timber & shade.
mkole	*Grewia similis*	Firewood, building poles, medicine, live stand for oyster nuts.
***mkichikichi/* msegese**	*Piliostigma thonningii*	Firewood, timber, medicine, shade, mulch, soil conservation, rope.
mkuzu	Unknown	Soil conservation, soil improvement, water retention.
mkeakiindi	Unknown	Firewood.
muhagati	*Olea europaea subsp. Africana*	Timber.
mbwewe	*Lecaniodiscus fraxinifolius*	Firewood, building poles.

Source: Author's fieldwork 1994-1998.

Table 3.12 Tree species planted prior to TFCG intervention

Vernacular Name	Tree Species	Indigenous or Exotic	Returns
mjohoro	*Senna siamea*	Exotic	Shade, building poles, firewood, ornamental, soil conservation.
mkungu	*Terminalia catappa*	Exotic	Timber, shade, soil conservation.
mkabela	*Grevillea robusta*	Exotic	Timber, building poles, firewood, soil conservation.
tiki	*Tectona grandis*	Exotic	Timber, building poles, firewood.
mchungwa	*Citrus sinensis*	Exotic	Orange fruit.
mwembe	*Mangifera indica*	Exotic	Mango fruit, firewood, timber.
paipai	*Garica papaya*	Exotic	Papaya fruit.
mlimao	*Citrus limon*	Exotic	Lemon fruit.
mdimu	*Citrus aurantifolia*	Exotic	Lime fruit.
mvugha	*Cedrella odorata*	Exotic	Timber.
mwaorubaini	*Azadirachta indica*	Exotic	Shade, medicine, soil conservation.
mrangaranga/ **mparachichi**	*Persea americana*	Exotic	Avocado fruit.

Source: Author's fieldwork 1994-1998.

were defined:

- Underlying causes represent the most fundamental level in explaining why there exists a need to grow trees;
- Returns from tree resources via tree products and tree services are more or less utilitarian reasons that farmers gave for growing trees; and
- Triggering factors, were isolated events or ideas that had provided the necessary impetus for growing trees.

The underlying cause for tree growing was villagers' concern for a reduction in availability of TFP, either due to reduced access to forest resources through reservation or due to a decline in preferred tree species due to over use, both in forest reserves and public forest. Some villagers in the area feared that the forest reserve boundaries would be extended,

Table 3.13 Tree species after TFCG intervention

Vernacular Name	Species	Indigenous or Exotic	Returns
tiki	*Tectona grandis*	Exotic	Timber, building poles, firewood.
mlimao	*Citrus limon*	Exotic	Lemon fruit.
mchungwa	*Citrus sinensis*	Exotic	Orange fruit, firewood.
***mrangaranga*/ mparachichi**	*Persea americana*	Exotic	Avocado fruit.
mkabela	*Grevillea robusta*	Exotic	Timber, building poles, firewood, soil conservation.
mwaorubaini	*Azadirachta indica*	Exotic	Shade, medicine, soil conservation.
mnazi	*Cocos nucifera*	Exotic	Coconut, alcohol, palm fronds.
cedrella	*Cedrella odorata*	Exotic	Timber.
mshai	*Albizia schimperana*	Indigenous	Timber, shade, soil conservation, nitrogen fixation.
mpingo	*Dalbergia melanoxylon*	Indigenous	Timber, carving.
mbambakofi	*Afzelia quanzensis*	Indigenous	Timber.
mtalawanda	*Markhamia hildebrandtii*	Indigenous	Building poles, firewood.
mkarambati	*Brachylaena hutchinsii*	Indigenous	Timber, flooring.
mpera mwitu	*Combretum schumannii*	Indigenous	Firewood, timber, tool handles, carving, medicine, mulch.
mvule	*Millicia excelsa*	Indigenous	Timber, shade.
mbwewe	*Lecaniodiscus fraxinifolius*	Indigenous	Timber

Source: Author's fieldwork 1994-1998.

further decreasing access to TFPs, particularly polewood and timber. One Kweboha villager, Mzee Shekerage (Box 3.8) felt that the demand for timber and polewood would increase in the future, both in the villages and towns. He hoped that by being one of the first to plant trees in the area, he would be one of the first to benefit in the future, by selling poles and timber to what he felt would be an increasing market.

Box 3.8 Mzee Shekerage of Kweboha, a sub-village of Kambai

Mzee Shekerage originates from Bumbuli in West Usambara. He was a pitsawer and a driver and had therefore learnt of different areas in East Usambara. He first moved to the Kambai area in the 1970s when he was given a contract to supply food to Muheza and Maramba hospitals. Later he decided to settle in Kweboha and farm for himself. His farm is approximately ten hectares and has very few unplanted trees, because in the past the land had been cleared for the Sigi-Miembeni Sisal Estate.

Shekerage has planted many trees on his farm: oranges; two hectares of teak (*Tectona grandis*); half a hectare of ***mkabela*** (*Grevillea robusta*) interplanted with maize; half a hectare of indigenous tree species such as, ***mpingo*** (*Dalbergia melanoxylon*), ***mtalawanda*** (*Markhamia hildebrandtii*), ***mshai*** (*Albizia schimperana*), and **mbambakofi** (*Afzelia quanzensis*); **mkuyu** (*Ficus sycomorus*) next to the river; and several of these species around his home for shade. He has already made money from half a hectare of teak (*Tectona grandis*), which he planted in 1989. It was thinned and each pole was sold for 300 TSh (40 pence sterling). This price is very low, since if he had his own transport he could sell them for 1,000 TSh each in Tanga.

Many people have been encouraged to plant trees after seeing the example of Mzee Shekerage. He doesn't see the increased numbers of people planting as competition. He hopes more will plant, because the more people with timber for sale the more vehicles will come in to Kambai to buy timber and that will help him also.

Source: Author's fieldwork 1994-1998.

Trees fill many different needs and are rarely grown for a single purpose only. The most frequently mentioned utilitarian reason for tree growing was the domestic need for construction timber, for instance, saw timber and poles. Villagers realise that timber requirements in the future will have to derive from cultivated trees. Although these trees may offer the services of shade and soil conservation, whilst offering other products such as firewood, fruits, medicine and rope.

Another major reason is the expectation of tree crops to generate cash income. Villagers feel confident that mature timber trees will generate a good price in the future. Fruit trees especially oranges and coconuts are being planted and managed as cash crops. Mature timber trees represent savings or assets for contingencies and unexpected needs. This aspect was a contingent in most individual tree growing strategies, whether they were wealthier villagers who could afford the land, labour, time and money to plant up woodlots (Box 3.9) or those that had planted several individual trees that they hoped would help them in their old age (Box 3.10 and Box 3.11).

Box 3.9 Christopher and Agnes of Kambai

Christopher and Agnes married in 1986. Christopher had a small farm that he extended by clearing forest and bush around his farm. His farm is now approximately six hectares. They employ eight casual labourers at the busiest times of year. This allows Christopher to live in Tanga and run his timber yard, whilst Agnes and her father manage the farm.

When clearing the land they retained trees on the hillsides and hilltops and a few trees for shade in the fields. He has planted oranges, coconuts, two hectares of teak (*Tectona grandis*), and interplanted ***mkabela*** (*Grevillea robusta*) with two hectares of maize. He first planted teak (*Tectona grandis*) four years ago after buying teak for his timber yard from Longuza Teak Plantation. He saw that the profit to be made was large and so thought that if he could grow it himself he could make a lot of money. Similarly, he spoke to TFCG about the soil conservation value of interplanting ***mkabela*** (*Grevillea robusta*) with maize and knew that it was a fast growing timber tree also.

In the future, he hopes to start his own tree nursery and employ one woman to manage it. He also wishes to plant many more teak (*Tectona grandis*) and ***mkabela*** (*Grevillea robusta*) trees. Other projects include starting a fishpond and irrigating his land by pumping water from the Muzi river by generator.

Source: Author's fieldwork 1994-1998.

Planting trees around farm boundaries has become increasingly popular in Kambai for securing land tenure and in assisting in boundary disputes with neighbouring farmers. Boundary planting was particularly popular with those without enough land to plant woodlots and were not convinced of either the merits of intercropping or the tree species utilised in intercropping.

Several reasons for tree growing concerning environmental protection were also given. They ranged from improving the micro-climate around the

Box 3.10 John Mponda of Kambai

John Mponda originates from Iringa, but came to the area when he was seventeen to work on the tea estates in Amani. After working for two years he saw that the work was very hard and difficult and he moved to Kambai in 1974 to farm. At first he found it difficult to obtain land, but eventually the elders advised him to clear an area of forest from SHUWIMU land. In 1975 he cleared three hectares of forest on which to cultivate.

When clearing the forest he retained some trees: ***mvule*** (*Millicia excelsa*) because it is a government reserved tree and good for timber; ***mshai*** (*Albizia schimperana*) for shade; ***mtalawanda*** (*Markhamia hildebrandtii*) for building poles; and many others next to the river to maintain the water source – something he is particularly aware of, being the village water pipe engineer. He has planted fruit trees such as, oranges, lemons, limes, mangoes and papaya and many bananas, which he has planted as windbreaks to protect his maize crops. He feels that planting and retaining trees is particularly important in preventing "hot winds from turning the land to desert". He commented that this is what had happened in his home village where farmers had cut down all the trees, which resulted in little rain and hot winds. He added that the government had forced people to plant trees and now there are many trees on farms in that area.

Five years ago he planted teak (*Tectona grandis*) on the boundaries of his fields. It had been the idea of his seven year old son who had seen others planting teak seedlings, which had been left over from the boundary marking of the forest reserves. His son brought the seedlings to him and asked him to plant them. In the future he wishes to plant more teak (*Tectona grandis*), ***mkabela*** (*Grevillea robusta*) and ***mkuyu*** (*Ficus sycomorus*) around his fishpond.

Source: Author's fieldwork 1994-1998.

home through shade to soil conservation and improving soil fertility.

Most individuals growing trees identified triggering factors that provided an impulse to start tree planting. The most common being when individuals had seen examples of tree growing, from a neighbouring farmer or well respected farmer. In Kambai the examples of Mzee Shekerage (Box 3.8) and Mzee Daili (Box 3.1) who had planted a one-hectare woodlot of teak (*Tectona grandis*) were often mentioned as examples that had been followed. Several farmers admitted to not knowing fully the benefits of

Box 3.11 Mama and Bwana Samora of Kweboha, a sub-village of Kambai

Mama Samora was the first in her family to become involved in tree growing, as she is Secretary of Kweboha women's group and leads the management of the group's teak nursery. Her husband has followed her lead by also starting a teak nursery, which he shares with his neighbour Muhammad. They plan to transfer half of their teak seedlings to their fields and plant as woodlots and to sell the remainder in the village. Upon harvesting his teak trees, he hopes to sell the timber for cash and plans:

- To build a modern house;
- To open a bank account; and
- To put money aside for his children to go to secondary school.

He feels that by planting trees his life will be easier when he is older. "Once growing in the fields they require little work in comparison to annual crops."

Source: Author's fieldwork 1994-1998.

tree planting, but having seen their friends and neighbours planting they followed suit not wanting to miss out on the perceived benefits. Kweboha's women's group was instrumental in spreading the knowledge of tree growing. Several men admitted that seeing the women involved in tree growing had given them the impetus to start themselves (Box 3.11). Others reacted to requests by their young children to plant trees after they had been given seedlings at school (Box 3.10).

Access to potted tree seedlings had also been a triggering factor. Friends and family living in close proximity to EUCADEP had given several villagers potted tree seedlings in the 1980s. With TFCG intervention, potted seedlings were widely available.

The choice of tree species to be planted was initially triggered by projects in the area. Kambai farmers' desire to plant teak (*Tectona grandis*) was a direct result of living in close proximity to Longuza Teak Plantation.

Tree products and services are not the only returns from forest land. Returns include both planted crops: for example, maize, cassava and bananas; and wild crops: for example, edible plants and fungi. These returns are described further in the following two sections.

Pattern of Returns from Forest, Field and Bushland Resources

This section again takes the case of Kambai village and analyses the local community's pattern of returns from forest, field and bushland derived resources. Table 3.14 shows the pattern of returns from land under different tenure regimes, as described by three sub-villages of Kambai village: Kambai, Miembeni and Msige. This table only represents those sub-villages of Kambai, Miembeni and Msige, since in producing the matrix villagers pointed out that those from Msakazi and Kweboha sub-villages were living and farming entirely on SHUWIMU land. Since these areas of SHUWIMU land have large areas of uncultivated bushland that has regenerated since the closure of Sigi-Miembeni Estate, these villagers are able to obtain 100 percent of their returns from SHUWIMU land. In contrast, Kambai, Miembeni and Msige villagers have access to only ten hectares of public forest and the remaining accessible land is under cultivation, with very little bushland. Even the Kambai, Miembeni and Msige villagers who are cultivating on what is officially SHUWIMU land have little access to bushland, because the majority of SHUWIMU land in close proximity to these sub-villages has been cleared for cultivation. These sub-villages have therefore required out of necessity to obtain illegally approximately a third of their returns from Kambai and Semdoe Forest Reserves (Table 3.14).

It is important to note the types of returns that are derived mostly from forest resources, namely medicine, fibres for ropes, meat and building poles (Table 3.14). The returns that are predominantly obtained from public land, which in the case of these sub-villages is predominantly fields under cultivation, are: fibres for roofing, vegetables and firewood (Table 3.14). Although Table 3.14 shows that timber is no longer obtained from any of the resources, which goes in line with the pitsawing ban, the author is aware of two separate cases of illegal pitsawing, which were conducted the year prior to the drawing of the matrix by villagers. Both had been illegally organised by the Government Forest Guard, who had felled one ***mvule*** (*Millicia excelsa*) tree on SHUWIMU land and two in Kambai Public Forest. The village government had reported these incidences to the EUCFP and the Forest Guard was forced to leave the village and was demoted to nursery worker elsewhere. Kambai village had also been granted a licence in 1995 to obtain timber from the public forest for reconstructing a bridge for the Kambai to Longuza road.

Whilst constructing the matrix a debate ensued as to whether villagers

Table 3.14 Pattern[14] of returns from land under different tenure regimes (Kambai, Miembeni & Msige sub-villages)

Returns	Land Tenure Regime: Central Government Forest Reserve Kambai & Semdoe	Public Land[15] (Field) Kambai	Private Land[16] (Field and Bushland) SHUWIMU
Medicine	///	/	//
Fibres for ropes	///	/	//
Fibres for Roofing	0	///	//
Vegetables	0	///	//
Meat	///	/	//
Firewood	0	///	/
Building poles	///	/	//
Timber	0	0	0
TOTAL	12	13	13
PERCENTAGE OF TOTAL	32	34	34

Source: Author's fieldwork 1994-1998.

were still illegally obtaining returns from the forest reserves. After some debate, an elder, Mzee Shekerage, summed up the discussion:

> In the past, we were getting all our needs from the forest, such as timber, building poles, firewood and ropes. Now there are very few people who are able to go to the forest to get ALL their own needs. The Forest Guards have remarked the boundaries of the reserves, so that everyone must know they are not allowed to disturb the forest. Although, it is difficult to find good

[14] Where /// denotes the majority of returns are obtained from that particular resource and / denotes the least.

[15] Kambai public land includes ten hectares of public forest, with the remainder being fields and bushland.

[16] Kambai, Miembeni, Msakazi and Kweboha villagers have cleared fields and farmed SHUWIMU land since the 1960s, although they are officially squatters. The vegetation cover is predominantly field and bushland.

quality building poles outside the forest, we can find alternative sources, by planting timber trees on our farms.

The pattern of returns from forest, field and bushland derived resources can be further analysed by examining individual household patterns of returns. Mary Christopher and her father Christopher Waziri of Kambai village drew a model of the flow of returns to and from her household. Table 3.15 is a representation of the model by the author, redrawn in order to show the relative dependency on forest, field and bushland derived resources. Table 3.15 shows the forest, field and bushland returns to the household of Mary Christopher. Mary's household consists of herself, her elderly father, her younger sisters' child, and her own five children. She is separated from her husband and receives no support from him for herself or his children. It is clear that her livelihood depends predominantly from subsistence returns, with few returns making cash. She makes approximately 18,000 TSh a year from her field by selling excess maize and sugarcane alcohol locally in the village. Cash is required for household necessities such as, salt, soap, occasionally tea and sugar, yeast and wheat flour, plastic buckets for carrying water, cooking pots and utensils; clothes; medicine; the maize milling machine; and paying village taxes. Often when one of Mary's children were ill and needed to go to the dispensary it would take her a week or two to collect enough money to take them. Her usual strategy would be to offer her services at the sugarcane juicer where she would be paid to prepare the sugarcane alcohol for the bar.

Her father who is too ill to cultivate contributes to the family income by making crab baskets and catching crabs and shrimps from the Miembeni River. His catch is not for sale but for domestic consumption.

Table 3.15 shows that the majority of returns come from the field, with fewer returns deriving from forest and bushland resources. Table 3.15 shows that present firewood and building pole requirements are obtained principally from forest resources. Mary and her father have increasingly become concerned about their future ability to obtain these returns from outside their fields and have therefore planted trees, which they hope to obtain timber and building poles from in the future.

Local Communities' Dependency on Forest Resources

Table 3.16 summarises the returns derived from forest resources. Local communities utilise a wide range of forest products for their daily subsistence (Table 3.16). As discussed, many of these products are not

Table 3.15 Forest, field and bushland returns to household: Mary Christopher of Kambai village

Resource	Type of Return		Return to Household Subsistence	Cash
Forest	Forest Product	NTFP	Edible fungi.	
		TFP	Firewood, building poles.	
	Forest Service	Water	Water, fish, crabs, shrimps.	
		Soil Conservation	Valued	
		Forest Land	Valued	
Field	Tree Products & Services	Retained	Shade, firewood.	
		Planted	Fruit (*Citrus sinensis, Cocos nucifera, Mangifera indica*). Timber, building poles (*Grevillea robusta, Tectona grandis, Afzelia quanzensis*).	
	Non-Tree Crops	Retained	Wild leafy veg., edible fungi.	
		Planted	Maize (12 sacks per year), Sugarcane, Bananas, Cassava.	Maize (2 sacks sold per year at 6,000 TSh each). Sugarcane (6 buckets of alcohol sold per year at 1,000 TSh each).
Bushland	Tree Products	Retained		
		Planted	Fruits,	
	Non-Tree Crops	Retained	Wild leafy veg.	

Source: Author's fieldwork 1994-1998.

only derived from forest resources, but are also obtained from field and bushland resources. Collection of forest products tends to be gender specific. Women are responsible for feeding the household and are therefore responsible for collecting edible plants and fungi, and firewood. Men are responsible for building the household and are therefore responsible for the collection of building poles, timber and fibres. Hunting, honey and medicinal plants are usually collected by specialists who tend to be men. Children are responsible for wild fruit collection. People are highly knowledgeable about the specific characteristics of different species of trees and plants and collection for use is species specific.

Table 3.16 Summary of returns derived from forest resources by local communities

Forest Products	Non-Timber Forest Products	**Wild foods**
		edible plants
		edible fungi
		fruit
		honey
		meat
		Fibres
		mats and baskets
		brushes
		thatch
		rope
		Medicinal plants
	Timber Forest Products	**Firewood**
		Building poles
		Timber
Forest Services	Forest Land	**Agricultural land**
	Ecological Services	**Water and soil conservation**

Source: Author's fieldwork 1994-1998.

NTFPs include wild foods, fibres and medicinal plants (Table 3.16). In East Usambara at present NTFPs are generally used for subsistence and are not greatly utilised commercially. Wild foods, whether derived from forest, bushland or field resources are utilised in all local households. Wild foods can be regarded as a socio-economic buffer for local communities that may be unable to obtain alternatives due to lack of cash and lack of access to alternative sources, such as for example, a butchers. Collection of edible fungi is seasonal, with the majority being collected after the rains. Some households depend on wild edible plants more frequently in the dry season when there may be a decrease in the availability of planted vegetables, which may coincide with a reduction in household income from the ***shamba***.

In areas where local communities have access to public forest resources, such as Magoroto Hill, women show a preference for the taste of forest derived edible plant species. Although these communities show a preference for forest derived edible plants, there is still an abundance of knowledge of field and bushland derived plants. Those without or with little access to forest resources collect predominantly field and bushland derived species.

Fibres for weaving baskets, making brushes and ropes predominantly derive from forest resources. Local communities have noted that the availability of these fibres has been reduced, through lack of access to forest resources and reduced abundance in the forests. Those without forest resources on their farms (Table 3.16) have to find alternatives. Fibres for weaving are increasingly bought from the local Muheza and Tanga markets and sisal is increasingly being used, although it is seen as inferior to forest derived ropes. Fibres for thatching are usually collected from bushland resources, although those that can afford to tend to prefer corrugated iron roofing, since it needs less maintenance.

Medicinal plants have not been highly commercialised in case study communities. Although some medicinal plants can only be derived from forest resources the majority can be found in field and bushland resources. If forest derived medicinal plants are required, they tend to be stolen from the forest.

TFPs include firewood, building poles and timber (Table 3.16). TFPs have customarily been derived from forest resources. However, access to forest resources has decreased with increased policing of forest reserves. Local communities have also noted that there is a reduction in the availability of good quality TFPs in both the public and reserved forests. This has led to the increasing realisation by local communities that TFPs will, in the future, have to be obtained from field and bushland resources.

Women prefer to collect firewood from forest resources if able, since there is likely to be a greater abundance of preferred tree species with desirable firewood characteristics. However, if a woman is old or ill and unable to go far to the forest then she will usually collect from fields and bushland. Similarly, those without access to forest resources must collect from field and bushland resources. Those who have retained or planted trees will have less difficulty in obtaining firewood in this way. In rural communities alternatives to firewood include kerosene, which wealthier households may use for cooking stoves. Charcoal, which is widely used in the urban areas, is not utilised in rural areas.

For the majority, building poles and timber derive almost exclusively from forest resources. Men complain, however, that the public forests no longer have enough good quality tree species for building poles and timber. They therefore must either steal polewood or timber from the forest reserves or obtain it from field and bushland derived resources. Individuals are beginning to specifically retain and plant trees in their fields for polewood and timber. Alternatives to polewood include mud bricks, however, timber is still required in construction and it is only wealthier families who can afford to pay experts for construction.

In conclusion, the collection of NTFPs derived from forest resources is in general a function of access, availability and preference rather than need. Alternatives are obtained from field, bushland and market sources. TFPs, particularly building poles and timber, can still be difficult to obtain from outside forest resources and are in fact even difficult to obtain from forest resources. Many individuals are well aware of this and are ensuring that they will have access to alternative sources by retaining and planting trees in fields and on bushland.

Forest services include the potential of forest land that is realised through its conversion to agricultural land and the ecological services that forest provides, such as water and soil conservation (Table 3.16). Forest-adjacent communities customarily obtained agricultural land by clearing forest. With the majority of remaining forest gazetted as forest reserves in East Usambara this potential is now lost to local communities. Communities alone and with assistance from government projects are increasingly realising that the present agricultural land will have to be managed more intensively in the future. They are also coming to the realisation that further deforestation could seriously jeopardise the ecological functions of the forest, such as water and soil conservation.

Chapter 4

Stakeholder Relationships, Rights and Responsibilities to Forest

Local Customary Era (1740-1892)

Usambara in the local customary era was a politically centralised kingdom, whereas Udzungwa was a series of separately ruled chiefdoms. The Usambara Kilindi king was **Ng'wenye Shi** (owner of the land) in Usambara (Feierman 1974). Ownership in this sense implied control over the land in its political aspect, in the same way the owner of a village (**Ng'wente Mzi)**, its patriarch, held rights over his progeny. To be owner of the land implied the right to take tribute from any of his subjects. Feierman (1974) demonstrates that for the Sambaa, there was a general structural relationship between ownership and the benevolent use of ritual power. In the local customary era, the Kilindi king held the ritual rain charms. It was believed that only if the wealth of the land - through tribute - was put into the hands of the Kilindi king, as though his own, would he 'heal the land' (**kuzifya shi**) by bringing rain. Without tribute, he would withhold the rain and would 'harm the land' (**kubana shi**). Hence, the Kilindi king was made owner of the land - in the political sense - for the collective good of his people - the public.

In East Usambara in the local customary era, land cover could be classified into different categories. Table 4.1 shows the terms that East Usambara elders believed were in use in the local customary era. Land cover in general was differentiated into two basic groups: wilderness areas such as, **mzitu**, **msitu** and **nyika**, which were uncultivated; and domesticated areas such as, **kaya**, **mzi, tanga** and **poli**, which were cultivated (Table 4.1).

The principal connection between local communities and wilderness was the belief in both Usambara and Udzungwa that wilderness held regenerative and healing powers, which were primarily associated with water or rain. These did not thrive in the domesticated sphere, but in wilderness. Hence, rituals for healing took place in the wilderness closest to home, the forests (**mzitu** and **msitu**). Note that the general term for 'forest' in use in the 1990s in both Kiswahili and Kisambaa is ***msitu***. However, as Table 4.1 shows, 'forest' was differentiated into two distinct types: **msitu** (open grassy forest) and **mzitu** (thick dense forest). These ritually important forest areas will be referred to as 'ritual forests'. Several forest areas, which have not specifically been connected to rituals, have also been connected to specific clans who were thought to have managed the area of forest. These clan-managed forests will be referred to as 'clan forests'. Key informants have identified six ritual forests and seven clan forests in the research area. Figures 4.1 and 4.2 locate the case study ritual and clan forests in East Usambara and Udzungwa respectively.

Table 4.1 Terms used for different land cover in East Usambara in the local customary era

Kiswahili	Kisambaa	English
mwitu	**mzitu**	Thick dense forest, large trees that stand close together, wilderness.
msitu	**msitu**	Open grassy forest, small trees and bushes, wilderness.
nyika	**nyika**	Bush and scrub, few trees, wilderness, lowland.
pori	**poli**	Grassy, open scrubland, no trees, bushland, fallow.
shamba	**tanga**	Field, garden plot, cultivated land.
kijiji	**mzi**	Village.
nyumba	**kaya**	Home.

Source: Author's fieldwork 1994-1998.

Tables 4.2 and 4.3 summarise respectively, stakeholders' relationships, and rights and responsibilities to forest in the local customary era. Table 4.4 summarises forest land tenure regimes in the local customary era. Tenure regimes are defined as "socially defined rules for access to resources and rules for resource use that define people's rights and responsibilities in relation to resources".

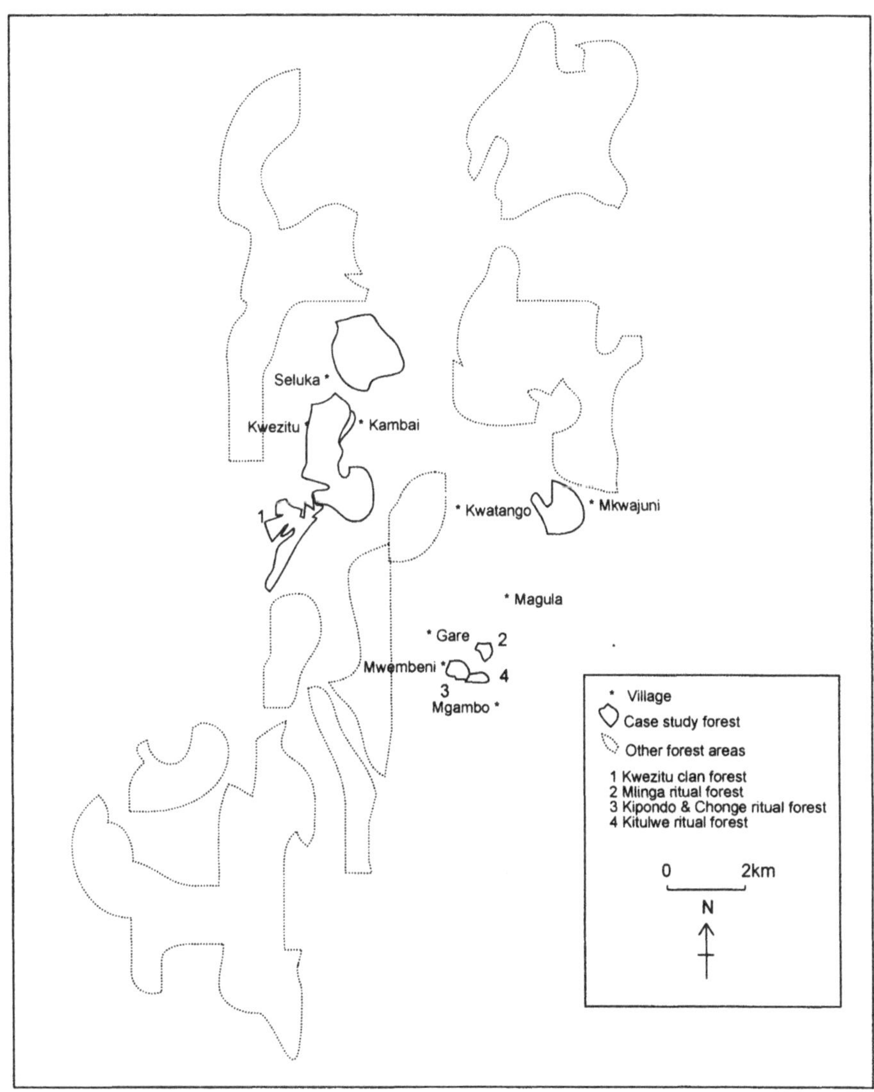

Source: Author's fieldwork 1994-1998.

Figure 4.1 Location of East Usambara case study ritual and clan forests in the customary era

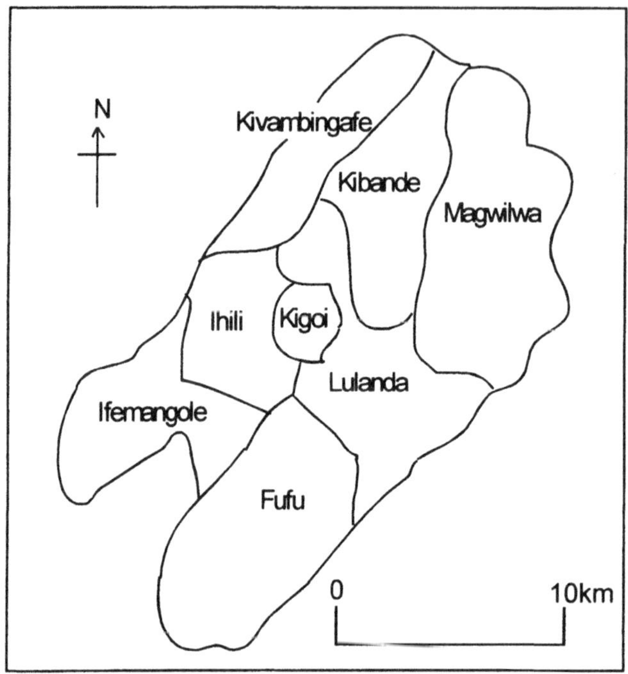

Source: Author's fieldwork 1994-1998.[17]

Figure 4.2 Sketch map of case study ritual and clan forests in Udzungwa in the local customary era

Ritual Forests (Mlinga,[18] *Kitulwe, Kipondo, Chonge, Kigoi, Fufu)*

Mlinga, Kitulwe, Kipondo, and Chonge are four ritual forests, which are situated on hilltops and ridges on Magoroto Hill in East Usambara (Figure 4.1). At each of these four forest sites there was a rain shrine where community leaders would make ritual sacrifices (***tambiko*****/fika**) to the ancestors in order to make rain, and 'heal the land' (**kuzifya shi**) from drought. It was customary for the Sambaa to bury leaders (Table 4.2) in

[17] From original sketch maps drawn by Lulanda elders.

[18] The name Mlinga in Kiswahili derives from ***linga*** which means to make equal, put side by side in order to match or compare, level, harmonise. The two peaks of Mlinga are side by side and almost equal in height. Since Mlinga was a ritual forest, it is apt that the name means to harmonise.

Table 4.2 Stakeholder relationships[19] to forest in the local customary era

	Forest status	
Stakeholder	Ritual forest (Mlinga, Kitulwe, Kipondo, Chonge, Kigoi, Fufu)	Clan forest (Kwezitu, Magwilwa, Lulanda, Itemang'ole, Ihili, Kibande and Kivambingafu)
Local community leaders	[C] Forest as an ambivalent place: leaders act as intermediaries between ancestral spirits and community members, where ritual sacrifice heals by ensuring protection from harm. Without ritual sacrifice the forest is malevolent and a placc to fear. [C] Forest as protector: a safe place to hide from enemies. [C] Forest as protector: a safe place to bury ancestors. [C] Forest as protector: a safe place to keep ritually dangerous medicines. [C] Forest as protector: a secret place for leaders to hold private meetings. [C] Forest as protector: a safe place to bury those with harmful or powerful spirits.	[C] Forest as an ambivalent place. [C] Forest as protector: a safe place to hide from enemies. [C] Forest as protector: of springs. [C] Forest as regenerator of life: provider of forest products and land.
Local community members	[C] Forest as an ambivalent place. [C] Forest as regenerator of life: provider of several forest products on the condition that forest rules are followed. [C] Forest as a healing place: traditional healers (***waganga***) collect medicinal plants and carry out ritual sacrifices for healing purposes. [C] Forest as protector: a safe place to bury those dead with harmful or powerful spirits. [C] Forest as protector: a place to carrying out initiation ceremonies in private.	[C] Forest as an ambivalent place. [C] Forest as protector: a safe place to hide from enemies. [C] Forest as protector: of springs. [C] Forest as regenerator of life: provider of forest products and land.

Source: Author's fieldwork 1994-1998.

[19] Where [C] denotes a customary relationship.

Table 4.3 Stakeholder rights and responsibilities[20] to forest in the local customary era

	Forest status	
Stakeholder	Ritual forest (Mlinga, Kitulwe, Kipondo, Chonge, Kigoi, Fufu)	Clan forest (Kwezitu, Magwilwa, Lulanda, Itemang'ole, Ihili, Kibande and Kivambingafu)
Local community leaders[21]	[C] Right and responsibility to make and uphold forest rules. [C] Right and responsibility to enforce sacrifices or fines on those not adhering to forest rules. [C] Right and responsibility to access and use forest for ritual sacrifice.	[C] Right and responsibility to make and uphold forest rules. [C] Right and responsibility to enforce sacrifices or fines on those not adhering to forest rules. [C] Right and responsibility to distribute clan land to clan members for cultivation. [C] Right and responsibility to permit or prevent the felling of trees. [C] Responsibility to carry out ritual sacrifices prior to the felling of a tree.
Local community members[22]	[C] Responsibility to adhere to forest rules. [C] In Udzungwa, no right to access forest. Forest was for leaders only. [C] In Usambara, right to access and use non-timber forest products, on the condition that only one product per person per trip was collected, and a sacrifice was left at the sacrificial stone.	[C] Responsibility to adhere to forest rules. [C] Right to access and use clan land. Responsibility not to transfer land outside clan. [C] Right and responsibility to fell trees with permission and ritual sacrifice from clan leaders. [C] Right to access and use forest products, but in Usambara, on the condition that only one product was collected per trip.

Source: Author's fieldwork 1994-1998.

[20] Where [C] denotes a customary right or responsibility.

[21] Local community leaders in the local customary era in East Usambara consisted of individual clan leaders and their clan elders, who were themselves led by village chiefs of the Kilindi clan and ruled over by the Kilindi king. In Udzungwa community leaders consisted of individual clan leaders and their clan members who followed the Hehe chiefs.

[22] Local community members in the local customary era included clan members and village members.

Table 4.4 Forest land tenure regimes[23] in the local customary era

	Individual	Group
Private[24]	[C] Leases on individual parts of clan forest (Kwezitu, Magwilwa, Lulanda, Itemang'ole, Ihili, Kibande and Kivambingafu).	[C] Clan forests (Kwezitu, Magwilwa, Lulanda, Itemang'ole, Ihili, Kibande and Kivambingafu). [C] Ritual forests (Mlinga, Kitulwe, Kipondo, Chonge, Kigoi and Fufu).
Public[25]		

Source: Author's fieldwork 1994-1998.

clusters of graves on these hilltops and ridges.[26] It was important that these rain shrines were situated in forests, since as Magoroto Hill key informants consistently stated "*rain-making requires the shade of forest.*" Hence, it follows that ritual forest was required for healing (Table 4.2).

The Kilindi leaders held overall authority of these ritual forests (Table 4.3), because they held the rain charms required for rainmaking. Other clan leaders were also required in ritual sacrifice and so too held access and user rights to the ritual forests for this purpose (Table 4.3). The local community leaders also held corresponding responsibilities for making and upholding forest rules (Table 4.3).

Mlinga was said to be the most powerful ritual forest for rain making in East Usambara. The original rain shrine was situated at the highest peak. If there was drought or the rains had not fallen at the correct time, then a group of community leaders[27] led by the rainmaker[28] would congregate at the peak. The rainmaker who held the special rain charms[29] would invoke

[23] Where [C] represents a customary tenure regime.

[24] Where 'private' refers to rights and responsibilities held by non-State entities.

[25] Where 'public' refers to rights and responsibilities held by State entities.

[26] It is customary to mark the four-corners of graves by planting trees. ***Mkaburi*/mkabuli** (*Plumeria rubra*), literally the grave tree, was common in the past for its ornamental flowers. In Kambai **mtalawanda** (*Markhamia hildebrandtii*), again an ornamental tree, is the preferred tree today.

[27] The community leaders would probably consist of the Kilindi chief who was usually the rainmaker and clan leaders and elders.

[28] The rainmaker would usually be the Kilindi chief who had been sent by the Kilindi king to rule over the community and had been given rain charms by the king (Feierman 1974 & 1990).

[29] The rain charms were said to be a mixture of herbs and water kept in a pot called a **kiza** (Feierman 1990).

the spirits of the ancestors,[30] by sitting on the sacrificial stone,[31] while clan leaders walked around him chanting: "*Stop sleeping, wake up, bring us rain!*" They would sacrifice a sheep or goat and often alcohol at the sacrificial stone, and then cook, eat and drink the offering as part of the rite. The rain was then said to fall immediately and soak the skins of these leaders as they returned from the forest to home.

Mlinga, Kitulwe, Kipondo and Chonge were also seen by local community members as providers of forest products for subsistence livelihoods, and hence regenerators of life (Table 4.2). Local community members held access and user rights to these ritual forests on the condition that they adhered to forest rules (Table 4.3). In the case of Mlinga ritual forest, collection of timber forest products and the clearing of forest land was prohibited (Table 4.3). Collection of non-timber forest products was allowed on the condition that only one product per person per trip was collected (Table 4.3). It was also essential to offer a sacrifice at the sacred stone (Table 4.3). A sacrifice could be in the form of food or simply a tree leaf[32] placed in the crevice of the stone. If an individual did not follow these rules, it was believed that they would be punished by the ancestors and would become blind or lost in the forest and would not be able to return home (Table 4.2).

Kigoi and Fufu are two ritual forests in Udzungwa (Figure 4.2). In contrast to East Usambara, the ritual forests of Kigoi and Fufu were strictly **Pakane** (for leaders only) or **Pa Mutwa** (for chiefs only). Local community members, other than chiefs or clan leaders, were prohibited from entering the ritual forests (Table 4.3) and would fear the malevolence of ancestral spirits if they did so (Table 4.2).

Kigoi was named after a very wide tree found at the centre of the forest that was itself called Kigoi.[33] Under this tree was a cave, inside which it was believed lived a giant. Clan leaders led by chief Mkwawa would sacrifice (***tambiko***) sheep or goats to this giant in order to make rain and 'heal the land' or make peace and 'heal the community' (Table 4.2).

Later the cave was used as a place for chiefs, clan leaders, elders and

---

[30] In the traditional Sambaa rain rites the spirit of the most recent ancestor would be invoked, but in the Kilindi rites each royal ancestral spirit would be invoked (Feierman 1990). In 1994, Sekiteke was the ancestor who was invoked in rain rites at Mlinga (Author's fieldwork 1994).

[31] The rainmaker is said to levitate above the sacrificial stone when he invokes the spirits of the ancestors.

[32] Informants felt that any leaf would be appropriate, as no particular tree species was favoured.

[33] The tree Kigoi still stands in what is now Fufu forest, although it is dead.

warriors to prepare for war and hide from warring tribes such as the Ngoni, Sangu and Ngindo. War medicines (**lihomelo** or **ngimo**) and weapons, such as, spears, axes, bows and arrows, and bush knives were hidden there. Local history also tells how chief Mkwawa hid in Kigoi in the seven-year war between the Hehe and the Germans (1891-1898). From that time 'Kigoi' meant 'hiding place', a place of protection (Table 4.2).

Inside Fufu, chief Mkwawa liked to rest under the shade of a very large tree when preparing for war. This tree became known as **Kisupo cha Mkwawa** (Shade of Mkwawa) and no one else was allowed to use that tree. In 1881 chief Mkwawa led the Hehe war against the Ngoni and many were killed on both sides. On returning to Fufu, Mkwawa hung the dead body of one of his best warriors, Kilufi, on the tree as a sacrifice to the ancestors. The body was left to rot and eventually only the skull remained. Locals decided to call the forest **Kibanga cha Mutwa** (literally Skull of a Chief). **Kibanga** is the name for 'skull' in Kihehe,[34] but ***fuvu*** is the name for 'skull' in Kiswahili. Lulanda key informants believe that the reason the forest is now called Fufu is that it is a mispronunciation of ***fuvu*** (skull). The forest was regarded as a ritual site for sacrifice (***tambiko***) for peace and the safe keeping of ritually dangerous war medicines (**lihomelo** or **ngimo**).[35] It was believed that the Ngoni could not attack them again or follow them to the healing, protection of the forest (Table 4.2).

Box 4.1 shows a traditional story concerning Fufu forest and demonstrates the fear and respect for the forest that local communities held. The story tells of a traveller who is afraid to pass through the forest because of the ambivalent powers of the ancestor Kilufi. By respecting Kilufi's powers, it appears his ancestors guide him through the forest to the safety of the village. The storyteller feels that the meaning of the story is to respect traditional ways of living and you will be protected.

There are other types of ritual forests which local people have spoken about, but have not specified a particular forest (Table 4.2). There were areas of forest where individual traditional healers (***waganga***) specifically collected their medicines and carried out healing sacrifices. Other forest areas were kept specifically as burial sites for the dead whose spirits the living feared. These forests were called **kitundu wantu** (wilderness for the dead). The forest was thought to contain these evil spirits and prevent them

---

[34] Kihehe is the Hehe tribal language.

[35] Feierman (1974) identified forests in Usambara for keeping ritually dangerous war medicines, known as **ghaso**. In East Usambara, there is a previously forested area, locally known as **Gasoi**, which was previously thought to have contained war medicines (Bildsten 1998: personal communication).

Box 4.1 Mizumu ya Kilufi (The magical power of Kilufi)

"During the old days, there was one traveller from Ikulumu going to Uhehe. It took him three weeks on the way. One day, after travelling for a long time, he ascended a steep slope. On descending, he reached a place known as Ifinga, which means grazing land. He did not give up; he had to climb the steep slope of Ifinga. Eventually at night, he reached the beginning of Fufu forest, which held the tree known as 'Shade of Mkwawa'.

He did not attempt to cross Fufu forest, because it was so dark and he was afraid of the magical powers of the ancestor Kilufi. He decided to sleep beside the path, but before sleeping he prayed for peace and protection.

Believe it or not, before putting his head on the ground he heard children crying and men ordering the women to carry their children and others to carry their luggage, in order to continue with their journey through the forest. The traveller realised that, the people were continuing their journey in the Fufu forest without fear. He awoke and started following the sound of their voices through the forest. He could not reach them, because it was too dark to go fast. He continued hearing the sound at a distance, and kept following the voices.

Without realising how, he discovered he had reached the edge of the forest and was safely in the village. To his surprise he heard the voices again, but this time they were far behind him, back in the forest. He went to the first house, where he was given accommodation. The next morning he continued with his journey."

"The meaning of this story is: wherever you go, you must respect the traditional ways of living."

Source: Author's fieldwork 1994-1998.[36]

from harming people. Other forest areas were used for initiation rites.

In summary, ritual forests in the local customary era were places of healing, protection and regeneration of life for forest-local communities (Table 4.2). Ritual forests are demonstrated to have been under private tenure regimes, where tenure was held privately by the group (Table 4.4). In the case of East Usambara ritual forests, although all land was publicly owned in its political sense by the Kilindi king (Table 4.4), tenure was in practice held privately by the community as a whole. In contrast, in

[36] See Meshack & Woodcock 1998, for Kiswahili version.

Udzungwa it was the community leaders who alone held tenure.

Clan Forests (Kwezitu, Magwilwa, Lulanda, Itemang'ole, Ihili, Kibande, Kivambingafu)

In the civil unrest in Usambara between 1855 and 1895, the relationship between local communities and the forest of Kwezitu (Figure 4.1) was that of a place of protection from rival clans (Table 4.2). For those outside the local community entering the forest for war or evil it was a place to fear. Vincent Chamungwana, an elder from Kwezitu village told the following oral history, which demonstrates this protective but ambivalent relationship:

> The first Sambaa to come to the area did so to hide from their enemies in the very big, thick forest (**mzitu**). They believed that if you hid in the forest and your enemies tried to follow you, the forest would protect you and your enemies would become lost in the forest. Later, these Sambaa settled in the area and called the area Kwemzitu.[37] The first people settled on a big rocky-forested escarpment near the present day sub-village of Mkalamo. They lived there so that they could easily check for enemies trying to attack the village. They would then throw stones from this rock at any enemies trying to enter the village.

Access and user rights to Kwezitu clan forest were given to all clan members in the local customary era (Table 4.3), for the provision of forest products and services (Table 4.2). Although the forest provided these products and services, the relationship was one of respect for the forests' ambivalence (Table 4.2). If a local community member wished to fell a tree, either for timber or the conversion of forest land to agricultural land, sacrificial rites were first to be made (Table 4.2). This was in order to 'cool the anger of the ancestors', whose spirits were thought to abide in forest trees (Table 4.2). User rights were regulated by clan leaders whose responsibility it was to make and uphold forest rules (Table 4.3). Box 4.2 contains a description of forest rules for the management of river resources as told by a Kwezitu elder, Vincent Chamungwana. It is interesting to note that the rules mirror those concerning use of resources in the ritual forest of Mlinga. Similar rules were given to the collection of other non-timber forest products.

---

[37] The '**Kwe**' means 'for, to or at' and '**mzitu**' means 'thick forest'. Hence, '**Kwemzitu**' means to the 'thick forest'. The village is now known as, 'Kwezitu' rather than 'Kwemzitu' - the 'm' being dropped in pronunciation.

Box 4.2 Rules concerning the use of river resources in Kwezitu clan forest

"The first inhabitants of the area saw a small river deep inside the forest, which they called Netondwe. It never ran dry and contained many fish, shrimps and crabs. If you went to catch fish, you must only catch fish. If you went to catch shrimps, you must only catch shrimps. If you went to catch crabs, you must only catch crabs. If you took both fish AND crabs, you would not see the way home."

Source: Author's fieldwork 1994-1998.

Forest resource use was therefore controlled by restricting access and user rights to that of one product per person per trip (Table 4.3). Access and user rights to timber forest products were controlled more specifically. Any clan member who wished to fell a tree was duty bound to first request permission from clan leaders who were both entitled and responsible for permitting or preventing the felling of trees in both clan forest and clan land in general (Table 4.3). The clan leaders were responsible for carrying out sacrificial rites in order to maintain the benevolence of ancestral spirits who were thought to abide in forest trees (Table 4.3). The individual who wished to fell the tree was responsible for providing the goat or sheep to be sacrificed.

Clan leaders had the right and responsibility to distribute parcels of clan forest to clan members for cultivation (Table 4.2 & 4.3). Among the Sambaa there was recognition of land 'ownership' in perpetuity. 'Ownership' of land was not dependent on use as with other tribes in Tanzania, such as the Gogo. Individual field and forest parcels could be bought or sold with the permission of clan leaders. However, what was being traded was not 'ownership' of the land, but the right to use and improve that land (Table 4.3). Clan women had access and user rights to clan land, on the condition that upon marriage they would relinquish their rights to access and use clan land. This was because upon marriage women became part of their husbands' clan and would then get access and user rights to that clan land. If divorced or widowed, women were able to return to their fathers' clan and would again have access and user rights to their fathers' clan land.

The clan forests of Magwilwa, Lulanda, Itemang'ole, Ihili, Kibande and Kivambingafu in Udzungwa are thought by key informers to have been named after clan leaders or traditional healers (***waganga***[38]) who lived in or had authority over those parts of forest (Table 4.3). All clan members had access and user rights to clan forests, and were responsible for respecting the authority of the clan leaders[39] (Table 4.3).

In summary, clan forests in the local customary era were perceived as ambivalent places of protection and regeneration of life, provided forest rules were respected (Table 4.2). Clan forests have been demonstrated to be under private tenure regimes, with rights and responsibilities held by the clan as whole (Table 4.4). However, individual field and forest parcels inside clan forests are analysed as being under private tenure, but with rights and responsibilities held by individual clan members (Table 4.4).

Tree Tenure Regimes

The relationships, rights and responsibilities to trees (tree tenure regimes) of forest-local communities are analysed separately from that of forests since they are often separable from forests. However, in East Usambara customarily private rights to land were strong (Table 4.4), therefore rights to trees were highly correlated to land rights.

Within the ritual forests where tenure was held privately by the local community as a group (Table 4.4), rights to use tree products, such as firewood, were also held by the community as a whole. These rights were however, regulated by rules governing their use (Table 4.3). The disposal of trees was however, prohibited by community leaders (Table 4.3).

Within the clan lands where tenure was held privately by both the clan as a group and individual clan members through leases (Table 4.4), rights to use tree products were correlated to land rights. Where individuals held strong tenure to parcels of land or forest their rights to own or inherit, plant, use and dispose of trees on that land were strong. This applied to both planted and retained trees. In the trade of land rights to forest or field parcels, compensation for labour and the results of labour on the land were customarily paid. Trees would increase the price paid for land parcels. The traditional means of exchange in land transactions were domestic animals, namely goats and sheep.

Where individual clan land was fallow, clan members held rights to use wild tree species, but not to dispose of them. Therefore, women held rights to collect firewood from fallow lands.

---

[38] ***Waganga*** is plural and ***mganga*** singular.

[39] Unfortunately key informers were unable to tell me if there had been rules in respect to access and user rights in the clan forests.

Tree tenure benefited those with stronger land tenure. Women whose land tenure depended on their relationship with men, either as daughters or wives, held rights to plant and use trees on land they had access to. However, upon leaving the clan through marriage, divorce or widowhood, they would relinquish their rights to access and use clan land, and hence their rights to use trees on that land even if they were the planters. Hence, although tree use and planting by women was not forbidden, their weaker land tenure did not encourage it.

Where the tenure of clan forest was privately held by the clan rather than individual clan members then rights to planted trees were held by the planter, whereas the rights to wild trees were held by the clan as a group. The right to decide to dispose of trees was however held by clan leaders (Table 4.3).

Local community leaders protected certain tree species whatever land tenure the land was under. **Mkuyu** (*Ficus sycomorus*) and **mvumo** (*Ficus thonningii*) were thought to have both malevolent and benevolent spirits living underneath these trees in springs. Disposal of the trees was prohibited, since protection of the tree, the spirits habitat, ensured that the spring would not run dry. It was prohibited to cut and burn the **mkanya**[40] tree. If it was burned, it was thought that a lion would enter the settlement. **Muungu** *(Cola usambarensis)* was not planted near the home, since it was believed to house evil spirits (***meshetani***) and would bring bad luck to the home and its inhabitants (Muir 1998: personal communication). In lowland areas, ***mbuyu***, the baobab, was used for sacrificial purposes.

Technocratic Era (1892-1989)

In the technocratic era all land was vested in the State. In the German administration the imperial decree of 1895 vested land in the German Empire. In the British administration all lands were declared to be 'public lands' and placed under the 'control' of the Governor for the benefit of the inhabitants. Since Independence all land in Tanzania is vested in the President on behalf of all citizens, and all that can be transferred, bought or sold are rights in land, not the land itself.

Forests are referred to and discussed in this section via the status they were given in the technocratic era, namely: Central Government Forest Reserves (CGFR),[41] Local Government Forest Reserves (LGFR),[42] forest

[40] Botanical name unknown.

[41] Known in the British administration as Territorial Forest Reserves.

on public land, and forest on private land. Figures 4.3 and 4.4 locate the case study forests in East Usambara and Udzungwa respectively.

Tables 4.5 and 4.6 summarise respectively stakeholders' relationships, and rights and responsibilities to the forests of East Usambara and Udzungwa in the technocratic era. Table 4.7 summarises forest land tenure regimes in the technocratic era. Although in its political sense all land was vested in the State, in practice, land tenure regimes were both public and private referring to rights and responsibilities held by both groups and individuals (Table 4.7). A dualistic tenure regime also existed with customary tenure coexisting with statutory tenure (Table 4.7).

Central Government Forest Reserve (Manga)

German administration Many early European travellers were full of enthusiasm for the potential value of Usambara forest resources (Table 4.5). Krapf (1858: cited in Iversen 1991) wrote that the forests "*...must be worth millions for ship building...*" and "*how many mills and factories could have been built along the many riverlets in this country*".

The first Usambara Forest Ordinance, which reserved forest, was passed as early as 1895 (Schabel 1990). Decisions regarding Usambara forests were made in Dar es Salaam in the Department of Natural Resources and Surveying on the basis of reports by European forestry 'experts' who began to assess the commercial worth of Tanzania's forests (Table 4.5). By 1913, 54,371 hectares of Usambara forest were reserved, with nine forest reserves in West Usambara and eight in East Usambara (Iversen 1991). In the whole Tanganyika territory 231 forest reserves were declared in the period between 1906 and 1914 (Grant 1924). The official policy was that the forest reserves should fill both the local and export timber needs (Table 4.5).

The Germans evacuated Usambara in 1916 due to invasion by the British in the First World War.[43] During the period of 1914 to 1920 there was "*...massive native encroachment of reserved areas in Usambara...*" (Grant 1924). This was probably partially due to local people escaping enlistment and crop loss by relocating in the more remote forests, whether they were forest reserves or forest on public land. Rodgers (1993) also points out that forest resources were probably required for emergency civil and military use. Local people would have taken advantage of freshly cleared forest for

[42] Known in the British administration as Native or Local Authority Forest Reserves.

[43] The First World War ran from 1914 to 1918. See Joelson (1928) for decription of British take over of Amani, in East Usambara.

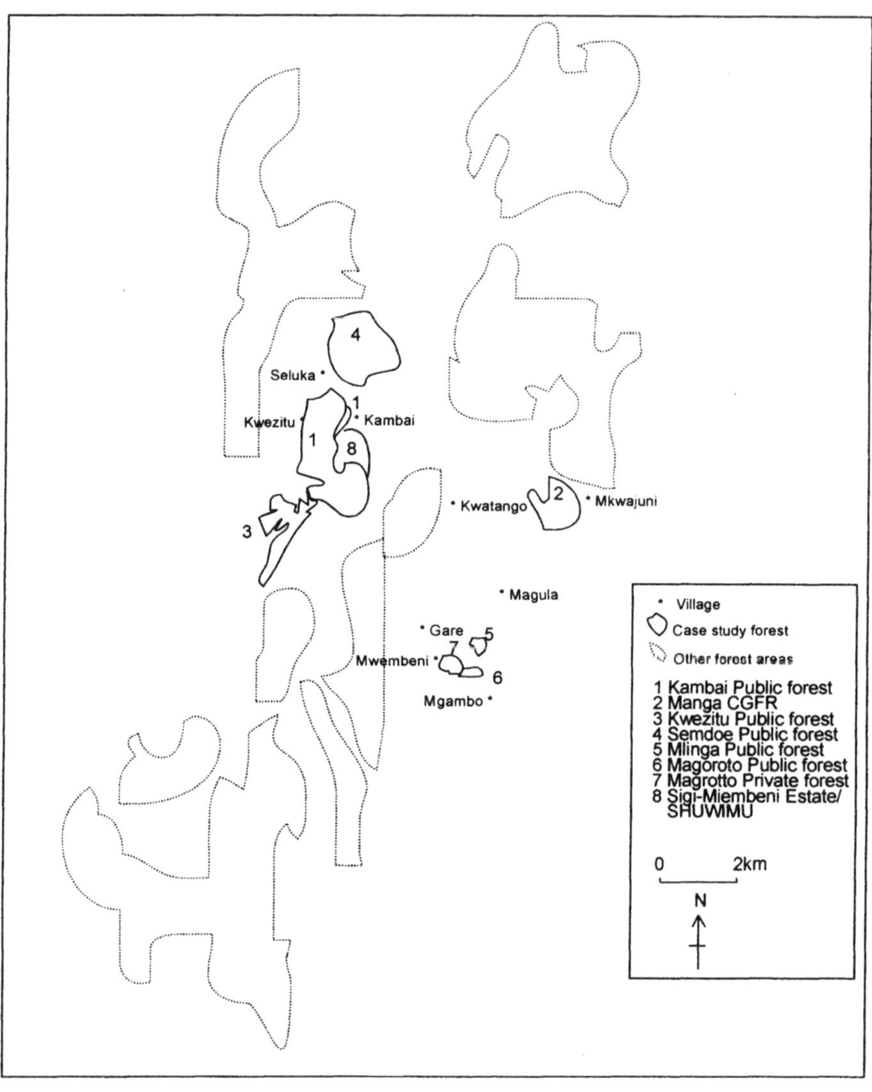

Source: Author's fieldwork 1994-1998.

Figure 4.3 Location of East Usambara case study forests in the technocratic era

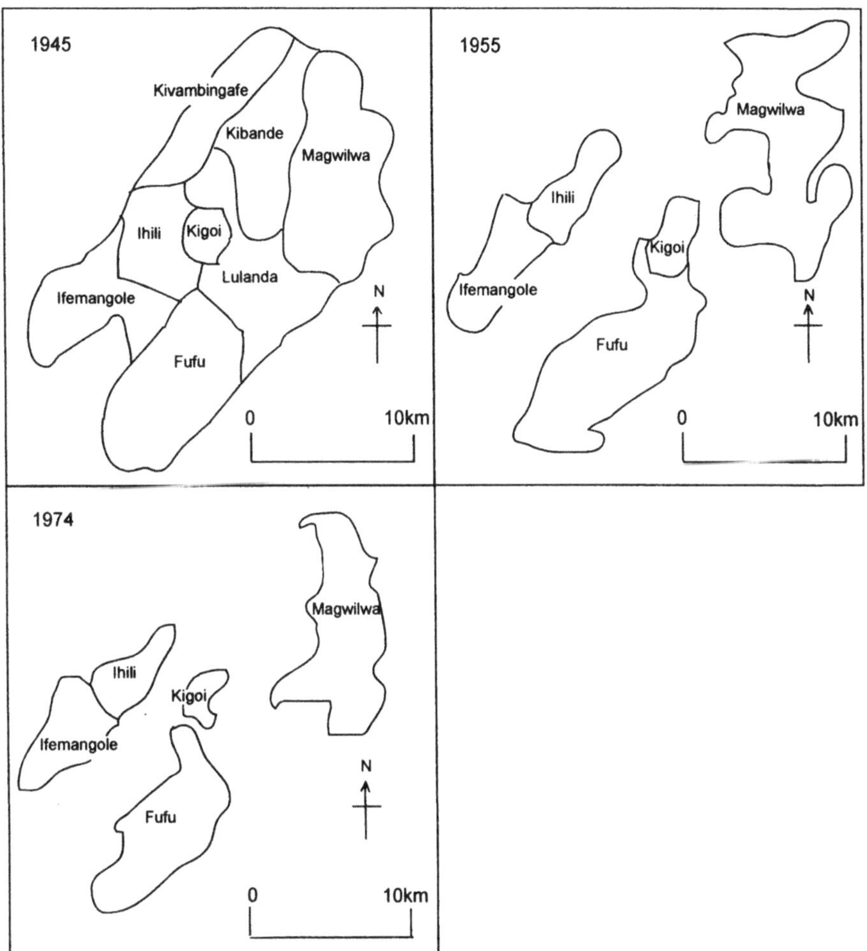

Source: Author's fieldwork 1994-1998.[44]

Figure 4.4 Sketch maps of Lulanda Local Government Forest Reserve: 1945, 1955 & 1974

agriculture. Conte (1996) argues that local people had learned to use European conceptions of land ownership as the basis for their own claims to land. For local communities whose relationship with the forest had been officially broken through reservation (Table 4.5), the escalation in deforestation was a reaction to the German forest policy of reservation.

[44] From original sketch maps drawn by Lulanda villagers.

Table 4.5 Stakeholder relationships[45] to forest in the technocratic era

	Forest status			
Stakeholder	Central Government Forest Reserve[46] (Manga)	Local Government Forest Reserve[47] (Lulanda)	Forest on public land (Kwezitu, Kambai, Mlinga, Semdoe, Magoroto Hill)	Forest on private land (Magrotto Estate & Sigi-Miembeni Estate/ SHUWIMU)
State[48]	[S] Forest valued for commercial worth. [S] Guardians of the forest. [S] Forest valued for ecological services.	[S] Forest valued for commercial worth. [S] Guardians of the forest.	[S] Forest resources for the benefit of local communities. [S] 'National trees' reserved by State for the potential value of timber returns.	[S] Forest valued for commercial worth. [S] In British administration, estates banned from destroying more forest.
Local community[49]	[S] Relationship broken. [C] Forest as provider of forest products and land. [C2] Forest reserve perceived as lost potential, therefore regain potential by converting to agricultural	[C] In German administration, relationship as in local customary era.[50] [S] Relationship broken in British administration. [C] [C2] Forest as provider of forest products and land.	[C] [S] Forest as provider of forest products and land. [C] In German administration relationship continued much the same as in local customary era.[51] [C2] In British administration customary relationship	[S] Squatters. [C2] Forest as provider of forest products and land.

[45] Where [C] denotes customary relationship developed in local customary era; [C2] denotes customary relationship developed in technocratic era; and [S] denotes statutory relationship.
[46] Known as Territorial Forest Reserves under the British administration.
[47] This classification of forest was created in the British administration. They were known as Native or Local Authority Forest Reserves.

	land.		gradually eroded. [C2] Forest as teaching ground for Makonde boys in initiation rites. [C2] Timber forest products valued for financial gain.	
Private Sector & International Community[52]	[S] Forest valued for commercial worth. [S] Botanical, amphibian, ornithological and catchment value research carried out by international scientists.	[S] Forest valued for commercial worth.	[S] Forest valued for timber returns.	[S] Forest valued for commercial worth.

Source: Author's fieldwork 1994-1998.

Local people were reclaiming land that had lost potential by being reserved from them and regaining that potential by converting it to agricultural land (Table 4.5).

Research in East Usambara began in the late 1890s with substantial collections being undertaken and the Amani Botanical Gardens being initiated (Table 4.5).

[48] In the technocratic era the State went from the colonial administrations of the Germans and British to be an Independent State (Chapter Two).

[49] In the technocratic era, the local community consisted of all village members, represented after Independence by the Village Assembly (Chapter Two).

[50] See Table 4.2 for details of relationships between local communities and forest in local customary era.

[51] See Table 4.2 for details of relationships between local communities and forest in local customary era.

[52] The private sector includes estates and sawmills and the international community includes researchers and donor funders.

Table 4.6 Stakeholder rights and responsibilities[53] to forest in the technocratic era

	Forest status			
Stakeholder	Central Government Forest Reserve (Manga)	Local Government Forest Reserve (Lulanda)	Forest on public land (Kwezitu, Kambai, Mlinga, Semdoe, Magoroto Hill)	Forest on private land (Magrotto Estate, Sigi-Miembeni Estate/ SHUWIMU[54])
State[55]	[S] Right and responsibility to control forest access and use. [S] Responsibility to employ staff to guard forest from illegal access and use.	[S] Right and responsibility to control forest access and use. [S] Responsibility to employ staff to guard forest from illegal access and use.	[S] Right and responsibility to permit or prevent the felling of government protected timber trees.	[S] Right and responsibility to permit or prevent the felling of government protected timber trees. [S] Right and responsibility to demand forest management plans for forest protected on private lands.
Local community[56]	[S] No rights. Responsibility to adhere to rules. [C2] Access and user rights to forest	[S] No rights. Responsibility to adhere to rules. [C] In German administration, rights and	[S] Access and user rights to forest products. [S] Responsibility to obtain	[S] No rights. Responsibility to stay out of forest. [C2] Right to access and use forest on estate

[53] Where[C] denotes customary rights and responsibilities; [C2] denotes customary rights and responsibilities evolved in the technocratic era; and [S] denotes statutory rights and responsibilities.

[54] Note that Sigi-Miembeni Estate was after Nationalisation in 1971 leased to SHUWIMU, the parastatal arm of Muheza District authorities, and so became theoretically under public tenure regime in that rights to the land were given to a State entity.

[55] In the technocratic era the State went from the colonial administrations of the Germans and British to be an Independent State.

[56] In the technocratic era, the local community consisted of all village members, represented after Independence by the Village Assembly.

	products, without corresponding responsibilities	responsibilities as in local customary era. [C2] In British administration, customary rights and responsibilities eroded. [C2] Access and user rights to forest products and forest land, by bribing forestry officials.	pitsawing permits from State. [C] Prior to 1940s and 1950s rights and responsibilities as in local customary era. [C2] In 1940s and 1950s local customary rights and responsibilities eroded. Right to access and use forest, without corresponding responsibilities	land that has been unused.
Private Sector and International Community[57]	[S] Rights to and responsibility to obtain logging licences.	[S] Rights to and responsibility to obtain logging licences.	[S] Rights to and responsibility to obtain logging licences.	[S] In German administration, colonists claimed rights to clear forest land for plantation estates. [S] In British administration, colonists responsible for managing forest land within estates.

Source: Author's fieldwork 1994-1998.

57 The private sector includes estates and sawmills and the international community includes researchers and donor funders.

Table 4.7 Forest land tenure regimes[58] in the technocratic era

	Individual	Group
Private[59]	[S] Estates where individuals were granted right of occupancy (Magrotto Estate and Sigi-Miembeni Estate). [C] Clan forest, where individuals were leased parcels of forest.	[S] Village Land where the Village has been granted rights of occupancy through Village Title Deeds. [S] Customary Land where a community has been deemed rights of occupancy. [C] Clan forests (Kwezitu, Magwilwa, Lulanda, Itemang'ole, Ihili, Kibande and Kivambingafu). [C] Ritual forests (Mlinga, Kitulwe, Kipondo, Chonge, Fufu and Kigoi.
Public[60]	[S] Central Government Forest Reserve (Manga). [S] Local Government Forest Reserve (Lulanda). [S] Leased right of occupancy (SHUWIMU).	[S] Forest on public land (Kambai, Kwezitu, Semdoe, Mlinga and Magoroto Hill).

Source: Author's fieldwork 1994-1998.

British administration To relieve the pressure on the forests, the British administration in 1920 initiated a Forest Department with a staff of 11 European foresters and about 100 local guards (Grant 1924; Troup 1936). All existing Forest Reserves were proclaimed anew in the Forest Ordinance of 1921. Gazettement continued and by 1942 the reserved area had about doubled (Iversen 1991). Forest officers justified the reservation of more forest by stating that the forest resources were far in excess of the local demand, and that this probably always would be the situation (Grant 1924). The Forest Rules of 1933 with later amendments regulated all forest activities for 20 years (Troup 1936).

Reserved forest, as in the German administration, was perceived

58 Where [C] denotes customary tenure; and [S] denotes statutory tenure.
59 Where 'private' refers to rights and responsibilities held by non-State entities.
60 Where 'public' refers to rights and responsibilities held by State entities.

predominantly in terms of its commercial worth (Table 4.5). As had its German predecessor, the British Forestry Department sought to replant newly cleared areas with fast growing exotics rather than granting access to local people. Usambara timber was needed to meet colony-wide demand for fuel, especially for the railways, building materials and high-grade exports. The Forest Department was charged, for example, with ensuring that Tanganyikan timber would be cheaper than imported Burmese teak (Iversen 1991). The British administration considered the management of valuable natural resources to be the exclusive domain of the State. R. S. Troup, the Director of the Imperial Forestry Institute from 1924 to 1939, maintained that: "*African forest use, which threatens the entire colony's timber supplies, is unjustifiable.*"

Forest reservations were often backed by conservation arguments, but in practice, forestry business and export interests prevailed in the management of the reserves (Table 4.5). However, the need to conserve forest (Table 4.5) and the value of forests to local people were more clearly acknowledged in a later statement of the official forest policy (Legislative Council 1953):

> Forests must be preserved in perpetuity in the public interest on behalf of the community as a whole. To demarcate and reserve in perpetuity, for the benefit of present and future inhabitants of the country, sufficient forested land or land capable of afforestation to preserve or improve local climates and water supplies, stabilise land which is liable to deterioration, and provide a sustained yield of forest produce of all kinds for internal use and also for export. To undertake and promote research and education in all branches of forestry and to build up by example and teaching a real understanding among the peoples of the country of the value of forests and forestry to them and to their descendants.

International scientists continued to be interested in biological research in East Usambara. In 1928, surveys were undertaken on amphibians and by the 1930s detailed ornithological work had begun (Table 4.5).

Manga was gazetted as a CGFR[61] by the British Forestry Department in 1955 (Government Notice 112: EUCFP 1995). The Forestry Department held the rights to access and use the forests for the benefit of the State (Table 4.6) and was responsible for employing forest officers and forest guards (Table 4.6). The forest officers and guards were in turn responsible for guarding against illegal access and use of the forests and imposing fines

[61] Known as Territorial Forest Reserves in the British administration.

on those not adhering to rules (Table 4.6).

Members of the forest-local community, Mkwajuni, held no statutory rights or responsibilities, apart from the responsibility to adhere to rules and stay out of the forest, locally known as Kwaboko[62] (Table 4.6). However, as was customary, the local community continued to obtain forest products (Table 4.6). In 1967 the eastern boundary of the reserve was moved back in order to give Mkwajuni villagers' 24 hectares for further settlement purposes.

Mlinga forest was proposed as a CGFR in 1954, but was not gazetted under the British administration.

Independent State The British-induced infrastructure regarding forestry continued largely unchanged after Independence. The commonly articulated principle of British administrative forest policies that "*...the satisfaction of the needs of the people must always take precedence to the collection of revenue...*" was retained (Government of Tanganyika 1945: cited in Wily 1997). However, in 1962, it was proposed that Manga CGFR be degazetted or converted into a teak plantation in order to collect revenue (Table 4.5). Although by 1967 it was still listed as a CGFR, and the eastern boundary of the reserve had been moved back 24 hectares to satisfy the Mkwajuni community's settlement needs (Table 4.3). Semdoe forest was proposed as a CGFR in 1964, but was not gazetted. Proposed Forest Reserves between 1974 and 1975, one of which was Kambai forest were also not gazetted (Kalaghe et al 1988; Hamilton & Bernsted-Smith 1989).

CGFRs became the responsibility of the Forest Division under the Ministry of Lands, Natural Resources and Tourism and in 1976 the Director of Forestry launched the Tanga Catchment Forest Project. The project included 25 natural forest reserves in Usambara, one of which was Manga, thereby also creating the category 'Catchment Forest Reserve' (Hermansen et al 1985; Lundgren 1985; Hamilton 1988). Biological research in the area increased steadily, with work in the area including an attempt to understand the drainage and catchment value of the forests (Bruen 1989; Litterick 1989) (Table 4.5).

Local Government Forest Reserve (Lulanda)

German administration In the German administration, local communities still held a customary relationship with the forests of what are now

---

[62] In 1994 the forest was still known locally as Kwaboko. Few villagers actually knew that the forest was officially named Manga by foresters.

collectively known as Lulanda forest. In the seven-year war between the Germans and the Hehe of 1891 to 1898, local history tells of how chief Mkwawa hid from the Germans in the protective forest of Kigoi.

British administration The British administration has generally been considered more attentive to local authorities than the German administration. For instance, two categories of forest reserve were erected: Central Government Forest Reserves (CGFR)[63] and Local Government Forest Reserves (LGFR),[64] the latter of which were managed by District authorities (Table 4.6) under the guidance of the Forest Department (Legislative Council 1953).

According to Lulanda villagers, it was around 1941 that officers from the Forest Department visited and marked the boundaries of Lulanda forest (Figure 4.7: 1945). The community of the time was asked to relocate to areas outside the forest reserve boundaries and were informed of the new status of the forest as Lulanda Forest Reserve. The present District Forest Officer believes that Lulanda was gazetted at this time as a LGFR, although no official declaration or boundary maps exist (Lovett & Pocs 1992). Whether the forest was actually gazetted or not the local community was led to believe that the forest was a State forest reserve. Their customary relationship with ritual and clan parts of the forest was officially broken (Table 4.5).

Local community elders tell how in the 1950s the British administration encouraged the community to return to the area on condition that they cultivate coffee. Elders suspect that the British used this as an experiment, and that they had plans to start coffee estates themselves, but in fact the British did not develop coffee estates. The community returned to the area and illegally cleared patches of forest for coffee and subsistence agriculture. Aerial photographs of Lulanda LGFR from 1955 show fields cleared inside forest areas. The gradual break up of the forest into patches through deforestation for agricultural land had begun. The local community's relationship with the forest had changed. The forest was now almost purely a provider of forest products and land (Table 4.5).

A Forest Attendant (FA) was placed in Lulanda in the 1950s and it was his responsibility to guard the forest as no access or use was permitted (Table 4.6). It was also his duty to fine those caught accessing or using forest resources (Table 4.6). In practice the FA allowed locals to access

---

[63] CGFRs were known as Territorial Forest Reserves in the British administration.

[64] LGFRs were known as Native or Local Authority Forest Reserves in the British administration.

and use forest products and convert forest to farmland on payment of bribes. Lulanda villagers say that the FAs neglect of his responsibilities went on in the full knowledge of the then District Commissioner (DC). This they felt was due to the fact that the FA was the middleman in the transport of ivory, which the DC would sell illegally. Customarily, the community leaders controlled access and use, but these rights and responsibilities were eroded with reservation of the forest and local control of the forest given to the FA (Table 4.6). With the FAs failure to carry out his responsibilities and abuse of power, the Lulanda community sought to access and use the forest whilst favourable conditions lasted and the forest remained open to all (Table 4.6). Lulanda forest, which had previously been under a customary private-group tenure regime, had moved to a statutory public tenure (Table 4.7) and in the process access to the forest had moved from closed to open.

Independent State Figure 4.6 shows clearly that gradual encroachment around deforested areas between 1955 and 1978, made the once contiguous forest patches into three discrete forest patches, with clear forest-field interfaces. In 1974 the Lulanda community had settled and registered Lulanda village in Villagisation, inside what was meant to be the boundary of Lulanda LGFR. Forest continued to be valued purely for its function of provider of forest products and land (Table 4.5).

In 1974 with Villagisation, the District authorities placed a Forest Attendant (FA) in the village. However, rather than protecting the forest, he illegally authorised pit sawing for timber and allowed villagers to obtain forest products on payment of bribes (Table 4.5). Villagers say that this went unquestioned by a senior government official, as he used the FA as his middleman in the illegal trade of ivory.[65] The degradation of the forest patches had begun, as locals perceived themselves to be thieves of State resources (Table 4.5).

Forest on Public Land (Kwezitu, Kambai, Mlinga, Semdoe, Magoroto Hill)

Throughout the technocratic era, forest that had not been reserved or taken as private estates, such as, Kwezitu, Kambai, Mlinga, Semdoe and Magoroto Hill forests (Figure 4.5), was officially given the status of forests on public land (Table 4.5). As far as the State was concerned this "*...remaining forest was for the benefit of the local people...*" (Legislative Council 1953) (Table 4.5). Although, those tree species valued for their

[65] Illegal because he did not go through proper legal procedures.

timber, such as ***mvule*** (*Millicia excelsa*) were reserved by the State as 'national trees' and local people required permits to fell such species even on agricultural land (Table 4.5). Local communities held statutory access and user rights to these forests, but without corresponding responsibilities (Table 4.6).

German administration In the local customary era, several of these case study forests had been ritual forests, such as, Mlinga and Kitulwe, Kipondo and Chonge forests of Magoroto Hill, and clan forests, such as, Kwezitu. Under the German administration, many of the customary relationships continued much the same as they had in the local customary era (Table 4.5).

For instance, the clan forest of Kwezitu continued to be a place for local people to hide from enemies, a place of protection. Elders tell of the harsh treatment local communities had under the German administration. Vincent Chamungwana, a Kwezitu elder tells how men and women were forced to build the road from Muheza in the lowlands to Amani in the highlands: "*Women were forced to give birth at the side of the road they were digging.*" It was at this time that Chamungwana feels that Kwezitu's population grew, with many families moving to Kwezitu - literally 'to the thick forest' - where they settled away from Amani and the German settlements (Table 4.5).

In the First World War the Germans took many Tanzanians as ***askari*** (soldiers) to fight against the British. Kambai local history tells of men who escaped the Germans as they were being transported at night to Tanga to fight. It is thought many of them passed through the forest of Kambai as they followed the sound of Zumbakuu. **Zumbakuu** - literally 'the great one' - is a waterfall on the Sigi River near what are now Kambai CGFR and Semdoe CGFR. It is said that as the water fell, it hit and entered a small opening to an underwater cave which created a large 'booming' sound, that is said to have been heard from Tanga![66] It was this sound that was thought to have guided the escapees through the forest at night. Once

[66] Mzee Mkufya, an elder of Kambai remembers a visit from a German dignitary residing in Amani:

> He sat in a special chair and was carried down the steep slope to Zumbakuu by eight men. On reaching the base of Zumbakuu he and his fellow Germans each took a bottle and one after the other threw it into the water. Many Europeans came and did this. I don't know why. We were afraid to ask in those days.

The 'boom' of Zumbakuu is said to have stopped, because the small under water cave was filled with bottles. One day, yellow sand poured out of the underwater cave and the 'booming' ceased.

again the forest protected the local community against their enemies (Table 4.5).

Local history also tells how the customary relationship with Mlinga ritual forest was eroded in the German administration (Table 4.5). Magoroto elders tell how a German visited the ritual forest of Mlinga in 1913. He lit a fire at the peak, which got out of control and set fire to the forest. Much of the forest was destroyed. Locals say that ancestral spirits were heard to cry, "*Where shall we stay now?*" since their tree abodes were destroyed. The German was said to die on his journey from the peak and this was said to be the work of the angry spirits. Since ancestral worship for rainmaking requires the shade of forest, it could not continue at the peak. The sacrificial stone was therefore moved from the peak and placed at the base of the peak where forest remained. The majority of the ritual forest was destroyed and devalued, but customary sacrificial rites continued in the much-reduced remaining forest (Table 4.5).

British administration Customarily, Mlinga and the forests of Magoroto Hill, namely Kitulwe, Kipondo, and Chonge had been managed as ritual forests and Kwezitu as a clan forest. Key informants in East Usambara believe that there was a distinct change in the relationship of local communities to their clan and ritual forests around the 1940s and 1950s (Table 4.5), when elders feel that customary tenure regimes were eroded (Table 4.6). An elder of Gonja, a sub-village of Kwezitu, recalls:

> It was some time in the 1950s that a man from Gonja went to the forest and cut a tree for his own uses without first asking the elders. When it was discovered what he had done he was forced by the elders to pay a fine of a goat.

Mzee Benjamin Maua of Bombo Village on Magoroto Hill believes that sacrificial rites for rainmaking in the ritual forest of Kitulwe were discontinued around the 1940s also. Locals believe decreased authority of community leaders to be the principal reason for the erosion of these customary relationships. However, customary relationships were known to continue in Mlinga ritual forest right into the 1990s.

In the British administration, many tea and sisal estates were in production, one of which was the Sigi-Miembeni Estate in the Kambai area. The Sambaa did not wish to work on these estates, after seeing that the work was hard with minimal returns. Hence, estate owners had to import labour from other areas. Many of these immigrant workers were Makonde from the Makonde Plateau in southern Tanzania and northern

Mozambique (Chapter Three: Box 3.1). These Makonde settled and brought with them their own beliefs and practices, some of which take place in the forests. One such practice is that of boys' initiation rites, which unlike the Sambaa rites of passage (Cory 1962a & 1962b), are held secretly inside the forest. For three months, boys as young as five go in a group to a secret place in the forest[67] with a Makonde teacher, where they are circumcised and learn life skills (Table 4.5).

Independent State Public forest was the only forest that local communities had statutory access to. Local communities therefore continued as was customary to obtain forest products and land from these areas (Table 4.5).

Logging which had been carried out in East Usambara since the German administration continued on a grander scale after Independence. The main logger was Sikh SawMills (SSM), first as a private company and then after nationalisation in 1971, as a parastatal organisation[68] (Table 4.5). From 1977, SSM was supported in its activities by FINNIDA - the Finnish Aid Agency (Table 4.5). As well as management advice, Finland provided equipment such as chainsaws, bulldozers, skidders for moving logs, and Finnish made logging trucks.

During the early 1980s, about eighty percent of the logs used by SSM were from sub-montane forests, such as Kwezitu and Magoroto Hill. Parts of Kambai and Semdoe forests were also logged. Logging was proceeding at the rate of one hectare a day and was directed at an 'intact' (unexploited or little exploited) forest where the desired tree species still remained (Hisham et al 1991).

Apart from SSM, the East Usambara forests also attracted individual pit-sawyers from other regions of Tanzania, especially Iringa and Mbeya (Table 4.5). Pit sawing involves the felling of individual trees, cross cutting the trunks and rolling the logs onto frameworks, usually over pits. Planks are then sawn off by hand with a large vertical saw operated by two workers, one above and one below the logs.

Kwezitu and Magoroto Hill communities benefited financially from

---

[67] This practice continues in Kambai village to date. However, I was unable to find out where or which forest the Makonde of Kambai use, but it is thought to be in forest on public land. Girls do not have their initiation until they are teenagers, and they stay in one house rather than going to the forest. Little is known about both the girls and the boys training, despite being very good friends with the girls' teacher for 1997. The Makonde initiation rites are still mysterious and secret and the Sambaa know little, but make up a lot! The coming out ceremonies of both girls and boys are however enjoyed by the whole community.

[68] Part of the Tanzania Wood Industries Corporation (TWICO).

logging and pit sawing (Table 4.5). The community was paid for each timber tree felled. For example, the Magoroto community bought an ISUZU lorry with the profits. The lorry was used for transporting crops to Muheza and Tanga markets. If pit-sawyers felled trees in customary clan forest then the clan or clan member would usually be paid compensation money for damage to crops and other trees caused by the pit sawing. They were also often given a number of planks of timber for their own domestic uses (Table 4.5).

Forest on Private Land (Magrotto Estate, Sigi-Miembeni/SHUWIMU)

In the technocratic era, colonists claimed large areas of forest for plantation estates (Table 4.6) and were granted right of occupancy (Table 4.7).

German administration German settlers first gained access to the interior in the 1890s and carved out large estates in forested areas (Meyer 1914: cited in Iversen 1991; Grant 1924; Forest Department of Tanganyika Territory 1930). Thesc areas were cleared for coffee, sisal, rubber, oil palm and teak commercial plantations (Table 4.5).

Magrotto[69] Estate on Magoroto Hill started in 1896 and forest was partially cleared for the commercial plantation of coffee (Table 4.5). Wohltmann (1902: cited in Iversen 1991) observed that Magrotto Estate, which was dealing in coffee still looked promising in 1898, but three years later the yields were declining like in all the other coffee estates and production changed to rubber.

British administration In 1921, Magrotto Estate's production changed to palm oil, with small amounts of black pepper, coffee and rubber. As the problems of soil erosion resulting from the earlier forest clearance became evident, plantation owners were banned from destroying more forest and the conservation of forests was seen as important (Table 4.5). Some plantation areas were declared official forest reserves, while others were replanted as forest reserves under compulsory purchase orders. The forests of several estates,[70] including Magrotto (Government Notice 99: EUCFP 1995), were given the status of forests protected on private lands.

[69] The Estate on Magoroto Hill is called Magrotto rather than Magoroto, which is the name of the area.

[70] Bulwa, Kwemtili, Monga, Msituni, Ndola, Ngua and Sangarawe were other estates in the East Usambaras that were given the status of forest protected on private lands.

Everybody possessing more than 200 hectares of forest was responsible for presenting forest management plans to the Government (Table 4.6). Estate managers were responsible for obtaining permits for the felling of government protected trees from the Forest Department (Table 4.6).

In the 1930s, approximately 800 hectares of forest in the Sigi Valley were cleared for the Sigi-Miembeni Sisal Estate (Table 4.5). The aerial photograph of 1954 shows the extent of deforestation. The estate consisted of a processing plant, estate worker houses, and a network of roads and electricity. The Decorticator[71] was brought from Mombassa in 1937.

Independent State In 1968, Magrotto Estate experimented with cardamom, cloves and tobacco and in 1987 hybrid palm oils were planted. The forests were still protected as forest on private land (Table 4.5).

Sigi-Miembeni Estate closed in the 1960s. Mzee Cosmos of Kambai village told of the events leading up to the closure of the estate:

> In the late 1950s, a Bishop visited Sigi-Miembeni and built a very big church. However, the manager of the Sisal Estate was a Muslim, and he burned the church down. On returning, the Bishop was very angry to see what had happened and he put a curse on the estate, saying that the land would never be successful in producing commercial crops again. Three months later, the estate manager died and parts of the estate were closed. Since this time, villagers believe that the land is cursed and no cash crops will thrive again.

After the closure of Sigi-Miembeni Estate and nationalisation in 1971, the land was leased by SHUWIMU,[72] the Muheza District Development Corporation, which is the parastatal wing of the district authorities. SHUWIMU continued to extract and process timber until the late 1960s (Tables 4.5 & 4.6). Since then, SHUWIMU have done nothing with the land, however, in 1996 there were newspaper reports to suggest that they were to start an orange plantation (Nipashe 1996), which never came to fruition.

Many of the immigrant workers of the defunct Sigi-Miembeni Sisal Estate and Sawmill settled and farmed in Kambai, or started the new sub-villages of Kweboha[73] and Msakazi,[74] both of which are on SHUWIMU

---

[71] A Decorticator is a machine that removes the fleshy part of the sisal plant from the fibre.

[72] SHUWIMU stands for SHirika la Uchumi la WIlaya ya Muheza (Muheza District Development Corporation).

[73] Kweboha means literally 'for alcohol'.

[74] Msakazi means literally 'for work'.

leased land. They held no official rights to the land, but many claimed old estate houses or built new houses and cleared farms inside land that was officially leased by SHUWIMU. Local communities, when convenient to them, had developed customary tenure regimes based on the example given to them by colonial administrations, namely that if land did not appear to be cultivated then it could be claimed through use (Table 4.6). Msakazi sub-village is on the site of the old Sigi-Miembeni Estate headquarters, where the majority of villagers live in old estate houses. Since both these sub-villages were and are still on SHUWIMU leased land, these communities are in theory squatter settlements, although informally the Village, Ward and District governments know and allow these communities to continue to settle and farm there. The regenerating woodland is seen as a provider of forest products and land (Table 4.5).

Tree Tenure Regimes

This section focuses on tree tenure in East Usambara[75] in the technocratic era. Tree tenure is analysed separately from forest land tenure, since rights to trees can be multiple and separable from land. For instance, the State held rights to all timber trees, such as ***mvule*** (*Millicia excelsa*) which were given the status of 'government protected or national trees', irrespective of the land tenure regime. The State held the rights to decide whether to dispose of such trees and was responsible for issuing permits for the felling of such trees.

In the case of non-government protected trees, tree tenure correlated with land tenure. So where local communities' statutory rights to land were weak, such as in forest reserves, private estates and publicly leased land, so were their statutory rights to trees. For as long as customary land tenure regimes coexisted with the statutory then so did customary tree tenure regimes. It was around the 1940s that both these regimes became eroded, when the authority of customary leaders in these areas was also eroded. Although where squatters had claimed land, customary tenure continued, such as in Msakazi and Kweboha sub-villages on SHUWIMU land.

Within public lands, local communities held statutory rights to inherit, plant, use and dispose of trees, as long as they were not government-protected trees. Customary rights to trees remained the same as in the customary era, although customary leaders' control on the disposal of trees was eroded.

---

[75] This part of the research focused on East Usambara rather than Udzungwa.

Participatory Era (1989-1998)

Forests are referred to and discussed via the official status they were given in the participatory era, namely: Central Government Forest Reserve (CGFR), Local Government Forest Reserve (LGFR), forest on public land, and forest on private land. Figures 4.5 and 4.6 locate case study forests of East Usambara and Udzungwa respectively in the participatory era.

In the participatory era - as in the technocratic - all land was vested in the President on behalf of all citizens, and all that could be transferred, bought or sold were rights in land, not the land itself. So in practice, land tenure regimes were both public and private referring to rights and responsibilities held by both groups and individuals. Similarly, as in the technocratic era, a dualistic tenure regime also existed with customary tenure coexisting with statutory tenure. Tables 4.8 and 4.9 summarise stakeholders' relationships, and rights and responsibilities to the forests of East Usambara and Udzungwa in the participatory era. Table 4.10 summarises forest land tenure regimes in the participatory era.

Central Government Forest Reserve (Manga, Mlinga, Kambai, Semdoe)

With an increase in awareness of the biodiversity and ecological value of the forests, the East Usambara Catchment Forest Project (EUCFP)[76] has worked with the mandate to protect the forest reserves of East Usambara since 1990 (Table 4.8). In phase one of the project, efforts focused on further gazettment and protection of forest reserves. Manga had been a reserve since 1955, but in the participatory era, Kambai, Semdoe and Mlinga forests (Figure 4.5) were gazetted as CGFRs: Kambai forest was gazetted in 1992, with the boundaries extended in 1993; Semdoe was proposed in 1993 and was gazetted in 1998; and Mlinga was gazetted in1994. In this way, although participation was the paradigm of the era, technocratic approaches of reservation were predominant.

Reservation, as in the technocratic era, increased the States rights and responsibilities in forest, whilst reducing the statutory rights and responsibilities of local communities (Table 4.9). In reserving forest on public land (non-reserved forest), statutory tenure regimes moved from public tenure where rights were held by the public as a whole - and hence local communities – to public tenure where rights were theoretically held

[76] EUCFP is implemented by the Forestry and Beekeeping Division (FBD) of the Ministry of Natural Resources and Tourism (MNRT) with financial support from the Government of Finland, and implementation support from the Finnish Forest and Park Service (FPS).

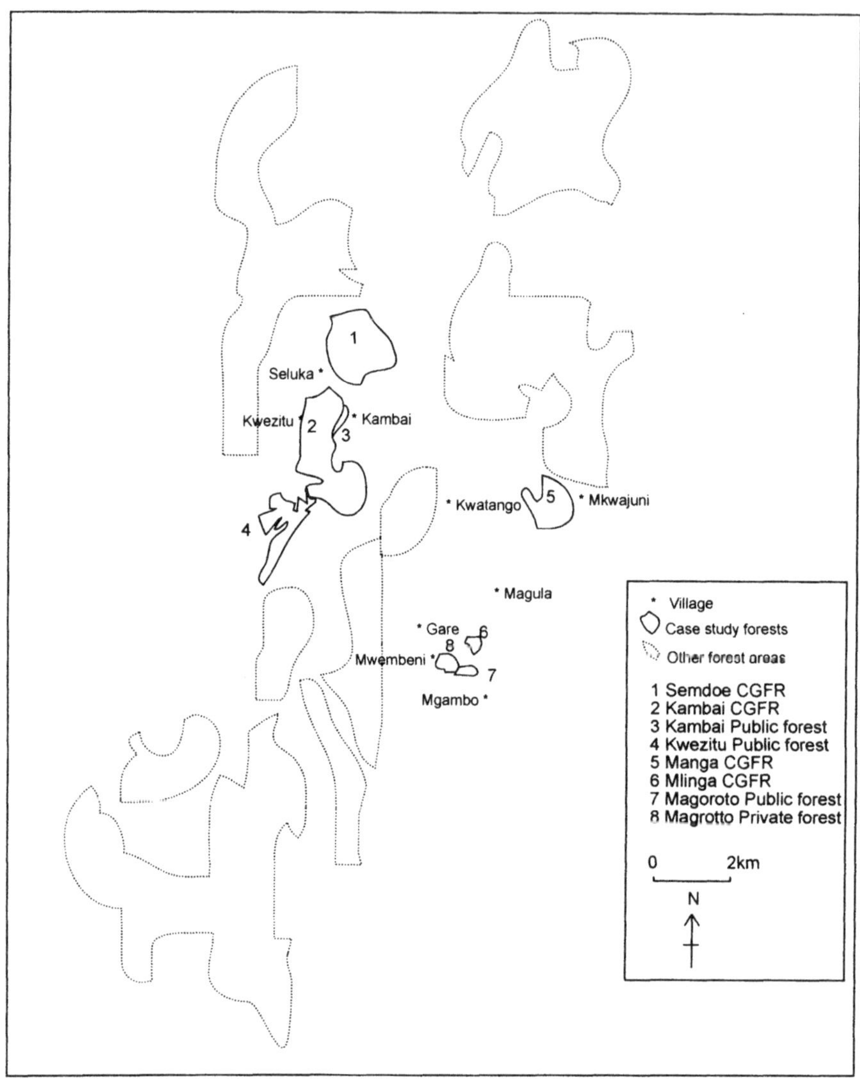

Source: Author's fieldwork 1994-1998.

Figure 4.5 Location of East Usambara case study forests in the participatory era

by an individual, the State (Table 4.10). The *de jure* tenure of the forests had moved from open to closed access. The *de facto* situation was however the opposite: the forests, which had in part been closed, had in fact become open.

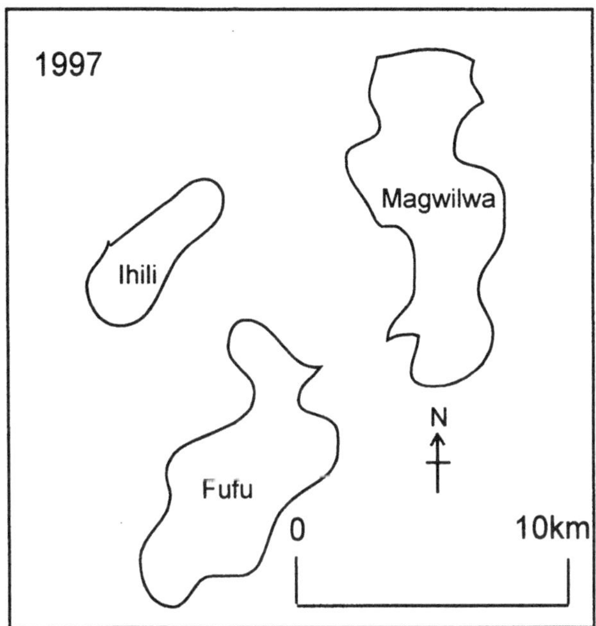

Source: Author's fieldwork 1994-1998.[77]

Figure 4.6 Sketch map of Lulanda LGFR in the participatory era

Although Mwembeni and Kambai villagers understood that it was illegal to enter or utilise Mlinga and Kambai forest reserves, many freely admitted to accessing and utilising the forest reserve for forest products[78] (Table 3.9 & Chapter Three). They bitterly referred to themselves, in relationship to the forest reserves, as "*mwizi-tu*!" - "only thieves!" - (Table 4.8). In this way, although officially management responsibilities had been taken over by the State, customary access and user rights continued to exist albeit illegally, but without corresponding customary responsibilities (Table 4.9). In the words of Benjamin Nantipi the Kambai Village Chairman: "*The forests are no longer ours, but theirs. The forests are out of our hands. We can only be thieves.*" By officially removing access and user rights from local communities, any responsibilities they may have felt towards the forest were also removed. All responsibility had been placed on the forest guard (Table 4.9).

[77] Sketch map from original maps drawn by Lulanda villagers.

[78] Women are permitted to collect dead wood for firewood in several forest reserves in East Usambara. Although it is not the case for any of these case study forest reserves.

Table 4.8 Stakeholder relationships[79] to forest in the participatory era

	Forest status			
Stakeholder	Central Government Forest Reserve (Manga, Mlinga, Kambai, Semdoe)	Local Government Forest Reserve (Lulanda)	Forest on public land (Kwezitu, Kambai, Magoroto Hill)	Forest on private land (Magrotto Estate and SHUWIMU)
State[80]	[S] Forest resources valued for biodiversity and ecological services. [S] Guardians of the forest.	[S] Forest resources valued for returns: timber and ecological services. [S] Guardians of the forest, but due to lack of funds efficiency reduced.	[S] Guardians of the forest for the benefit of local people. [S] 'National trees' reserved by State for the potential value of timber returns.	[S] 'National trees' reserved by State for the potential value of timber returns. [S] Guardians of the forest.
Local community[81]	[S] Relationship broken. [C3] The forest is perceived to be purely the domain of the State. [C3] Thieves of forest products and	[S] Relationship broken. [C3] The forest is perceived to be purely the domain of the State. [C] [C3] Users of customary	[C] [S] Forest as provider of forest products and ecological services. [C3] Forest perceived to be domain of the local community. Local	[S] Squatters. [C2] Forest as provider of forest products and land.

[79] Where [C] denotes customary relationship developed in local customary era; [C2] denotes customary relationship developed in technocratic era; [C3] denotes customary relationship developed or redeveloped in participatory era; and [S] denotes statutory relationship.

[80] The State is independent. EUCADEP and EUCFP are both State run projects, with financial and implementation support from international donors.

[81] The local community is synonymous with the village and the Village Assembly represents the village.

[82] Ulanzi is local alcohol made from bamboo sap.

	land.	pathway. [C] [C2] [C3] Forest as provider of forest products, particularly medicinal plants and ulanzi.[82]	community perceive themselves as guardians of the forest.	
Private Sector & International Community[83]	[S] Forest resources valued for biodiversity and ecological services. [S] Guardians of the forest.	[S] Forest resources valued for biodiversity and ecological services. [S] Guardians of the forest.	[S] Forest resources valued for ecological services. [S] Guardians of the forest.	[S] Forest resources valued for biodiversity and ecological services. [S] Guardians of the forest.
TFCG	[S] Forest resources valued for biodiversity and ecological services. [S] Guardians of the forest.	[S] Forest resources valued for biodiversity and ecological services. [S] Guardians of the forest.	[S] Forest resources valued for ecological services. [S] Guardians of the forest.	[S] Forest resources valued for biodiversity and ecological services. [S] Guardians of the forest.

Source: Author's fieldwork 1994-1998.

Villagers complained about the loss of access to forest resources and particularly agricultural land caused by gazettment and enlargement of reserves (Table 4.8). It is an idiosyncrasy that often villagers fields were incorporated into forest reserves, whilst forest areas sometimes were not. The EUCFP were required to pay villagers compensation for crops lost when fields were incorporated into the reserves. Some villagers continued to farm these fields, claiming delayed compensation as a loophole. For example, in 1998 two families still lived and farmed inside Semdoe CGFR. Several villagers tried to set deadlines for compensation, threatening further encroachment if the deadline was not met. As of 1998, many villagers had still not been fully compensated for crop losses.

---

[83] The private sector includes estates & the international community includes researchers & donor funders.

Table 4.9 Stakeholder rights and responsibilities[84] to forest in the participatory era

	Forest status			
Stakeholder	Central Government Forest Reserve (Manga, Mlinga, Kambai, Semdoe)	Local Government Forest Reserve (Lulanda)	Forest on public land (Kwezitu, Kambai, Magoroto Hill)	Forest on private land (Magrotto Estate and SHUWIMU)
State[85]	[S] Right and responsibility to control forest access and use. [S] Responsibility to employ staff to guard forest from illegal access and use. [S] Pitsawing banned.	[S] Right and responsibility to control forest access and use. [S] Responsibility to employ staff to guard forest from illegal access and use. [S] Pitsawing banned.	[S] Right and responsibility to permit or prevent the felling of government protected timber trees.	[S] Right and responsibility to permit or prevent the felling of government protected timber trees.
Local community[86]	[S] No rights. Responsibility to adhere to rules. [C2] [C3] Responsibilities eroded. Access and user rights to forest products,	[S] No rights. Responsibility to adhere to rules. [C3] Access and user rights for the collection of herbal medicines and ulanzi	[S] Access and user rights to forest products. [S] Responsibility to obtain pitsawing permits from State. [C2] [C3]	[S] No rights. Responsibility to stay out of forest. [C2] [C3] Right to access and use forest on estate land that had been unused for a

[84] Where [C] denotes customary relationship developed in local customary era; [C2] denotes customary relationship developed in technocratic era; [C3] denotes customary relationship developed or redeveloped in participatory era; and [S] denotes statutory relationship.

[85] The State is independent. EUCADEP and EUCFP are both State run projects, with financial and implementation support from international donors.

[86] The local community is synonymous with the village and the Village Assembly represents the village.

	without corresponding responsibility.	(bamboo alcohol). [C3] Responsible for assisting TFCG in labour of forest management.	Access and user rights, without corresponding responsibility. [C3] Local communities starting to take more responsibility for the management of forests, often with moral support from TFCG.	long period of time. [C3] District Commissioner informally encouraged Kambai Village government to motivate the sub-villages of Kweboha and Msakazi to continue cultivating and to start planting trees on SHUWIMU land.
Private Sector and International Community [87]	[S] Right to undertake research in the forests with research permits.	[S] Right to undertake research in the forests with research permits.	[S] No rights or responsibility.	[S] Right and responsibility to manage forests.
TFCG	[S] Right to undertake research in the forests with research permits.	[S] Right and responsibility to manage forest on behalf of the State.	[S] No rights or responsibility.	[S] No rights or responsibility.

Source: Author's fieldwork 1994-1998.

In Kambai, villagers saw both the advantages and disadvantages of forest conservation. Villagers frequently state the advantage of conserving forest for water catchment, a point that has been put across repeatedly in EUCFPs educational package. Although, villagers often feel that the disadvantages, in terms of loss of agricultural land and loss of potential agricultural land, are more important in the short term than conservation

[87] The private sector includes estates & the international community includes researchers & donor funders.

for water catchment (Table 4.8). One villager strongly disagreed with reserving Kambai forest and stated: "*I am very disappointed in the government. They are preventing my development. By conserving forest, they are leaving me without land.*" (Table 4.8). However, Mzee Shekerage, a Kambai elder said: "*...it is best to have foresters guarding the forest, because villagers here are only interested in making money and would destroy the whole forest without thinking*" (Table 4.8).

Research in the forests of East Usambara increased with greater international coverage caused by the large scale logging of the 1970s and 1980s and has highlighted the high biodiversity value of the forests and the Eastern Arc forests in general. Hence, at the same time as local communities' statutory rights to access and the use the forests were removed, increasing numbers of biological researchers - often foreign or Dar es Salaam based - were given access rights with permits to survey these same forests (Table 4.9). In July 1995 the EUCFP initiated and contracted forest biodiversity surveys for the forests of East Usambara. The surveys are conducted by Frontier-Tanzania[88] in collaboration with EUCFP. The aim of the surveys is to provide systematic baseline information on the biological values of different forests (Table 4.8) as a basis for management planning and long term monitoring, as well as training forestry staff in the use of biological inventory techniques. Of the case study forest reserves, Kambai, Manga, Semdoe and Mlinga forest reserves have been surveyed.

Organisations that are not working directly with the East Usambara forest reserves, such as EUCADEP and TFCG perceive themselves to be guardians of the forest (Table 4.8). By working in the public lands they attempt to contribute to the conservation of the forests by providing alternatives to forest products and improving land use management.

Local Government Forest Reserves (Lulanda)

In the early 1990s several biologists (Lovett & Congdon 1990; Lovett & Pocs 1992) surveyed (Table 4.9) Lulanda LGFR (Figure 4.6) and highlighted its high biodiversity value (Table 4.8) and its degradation. The three forest patches had a canopy up to 30 metres and were found to be intact in parts. Although, generally they were much disturbed following extraction of timber species in the technocratic era, opening the canopy in

[88] Frontier-Tanzania is a joint venture between the University of Dar es Salaam and the Society for Environmental Exploration, which is a British-based NGO.

Table 4.10 Forest land tenure regimes[89] in the participatory era

	Individual	Group
Private[90]	[S] Estates where individuals were granted right of occupancy (Magrotto Estate). [C] Clan forest, where individuals were leased parcels of forest.	[S] Village Land where the Village has been granted rights of occupancy through Village Title Deeds. [S] Customary Land where a community has been deemed rights of occupancy. [C] Clan forests (Kwezitu). [C] Ritual forests (Mlinga).
Public[91]	[S] Central Government Forest Reserve (Manga, Mlinga, Kambai, Semdoe). [S] Local Government Forest Reserve (Lulanda). [S] Leased right of occupancy (SHUWIMU).	[S] Forest on public land (Kambai, Kwezitu and Magoroto Hill).

Source: Author's fieldwork 1994-1998.

many areas. African mahogany, locally known as **mkangazi** (*Khaya nyasica*) and *Vitex amaniensis* had been extracted for timber in the technocratic era, but stocks had been exhausted. There was encroachment for cultivation along the edges of the forest and building poles, firewood and medicines were taken, which had continued since the technocratic era (Table 4.8).

As with CGFRs, all rights and responsibilities to Lulanda LGFR were held by the State, but in this case the District authorities (Table 4.9). The District authorities admitted however, to a reduced ability to manage the forest due to lack of funds. The fact that the Forest Attendant lived 12 kilometres away and rarely visited the forest and the District Forest Officer had not visited the forest for 12 years demonstrated their lack of ability to manage the forest. The efficiency of the District as guardians of the forest

[89] Where [C] denotes customary tenure; and [S] denotes statutory tenure.

[90] Where 'private' refers to rights and responsibilities held by non-State entities.

[91] Where 'public' refers to rights and responsibilities held by State entities.

was minimal (Table 4.8).

With increased awareness of the high biodiversity value of Lulanda LGFR, the Tanzania Forest Conservation Group (TFCG), a local NGO, started discussions with the governments of Lulanda Village and Mufindi District, concerning the initiation of a community-based forest conservation project in Lulanda. In September 1993, TFCG started the Lulanda Forest Conservation Project (LFCP). Box 4.3 lists the major project activities. Since 1993, TFCG assisted the District authorities to manage Lulanda LGFR using the predominantly technocratic approach of reservation and control (Table 4.9). Perceiving their role to be as guardians of the forest, in the reduced efficiency of the State (Table 4.8). Hence, the predominant project activities of physically managing the forest (Box 4.3). Other project activities included involving the local community in farm forestry, community woodlots and environmental education (Box 4.3), which were typical interventions of the participatory era.

Box 4.3 List of major LFCP activities

- Demarcating new forest boundaries;
- Planting boundaries with *Hakea saligna*;
- Making and maintaining fire lines;
- Planting and managing a forest corridor between two of the three forest blocks: Fufu and Magwilwa;
- Advising villagers on the management of a community woodlot;
- Advising villagers in farm forestry;
- Teaching environmental education in Lulanda and Mungeta (Isipii) primary schools and in the villages in general through sign boards, posters and village meetings; and
- Assisting village development by advising a womens' group in managing a maize mill.

Source: TFCG 1993-1998.

Despite TFCGs presence, the local community still illegally collected forest products and used forest land (Table 4.8). In 1996, Magwilwa forest patch was heavily burnt due to adjacent field clearing fires; a plot of tobacco was found growing inside the forest; and TFCG staff reported illegal timber extraction. On investigation eight pit sawing sites and over fifty planks of timber were found. The FA was found to be illegally authorising pit sawing. The Ward Executive Officer and Village Chairman confiscated the timbers and the DFO warned the FA not to do it again.

Following these events in 1996, TFCG stepped up their role as guardians of the forest (Table 4.8).

TFCG held a series of meetings with the local community in order to develop a better understanding of their relationship to the forest. In one such meeting where the management of the forest and the co-operation of the local community in assisting TFCG was discussed the overwhelming response was that: "*The right to be concerned about the forest was taken away when the forest was made a reserve. The forest is for the forest officer and his forest attendant.*" The local community felt strongly that their official relationship with the forest was completely broken (Table 4.8), and the only people who had benefited from the forest since it was made a LGFR were State employees. They could see no reason why they should assist the State in its work of management. Although this was the general feeling in the community, individuals did assist TFCG by donating land and transferring seedlings to the area to be planted up as a forest corridor. The reason they gave for assisting was that the forest was an important source of medicine and other forest products (Table 4.8) and it had always been there. They did not want to lose it entirely. The Lulanda Women's Group also agreed to assist TFCG, since TFCG had assisted them in obtaining a maize-milling machine. In this case, labour was being traded for material benefits. The typology of participation was participation for material incentives (Chapter One: Table 1.2).

Later still, the village, through the village government, agreed to assist TFCG in monitoring fires and illegal activities in the forest. The village had taken on management responsibilities, albeit willingly, without corresponding rights (Table 4.9). Later still TFCG were visited by a group of elders, who requested that they be permitted to access Lulanda forest for the collection of medicinal plants and **ulanzi** (bamboo alcohol) (Table 4.9). TFCG had already informally allowed villagers to access a customary pathway through Fufu forest patch, now they also allowed the continuation of the customary collection of medicines, honey and **ulanzi** to go unhindered (Table 4.8). User rights were being traded for local participation in the work of managing the forest.

Forest on Public land (Kwezitu, Kambai, Magoroto Hill)

With an increase in awareness of the biodiversity and ecological value of the forests, the East Usambara Conservation and Agricultural Development Project (EUCADEP)[92] started in 1989, with the mandate to work in the

---

[92] The project was implemented by the Ministry of Agriculture and Livestock, with financial

public lands. The principle was that agricultural development, through for example, farm forestry and improved land use management, would offer alternatives to forest products and reduce the need for forest land, thus assisting in the conservation of forest. In this way, both the State and the international community perceived their relationship to be that of guardians of the forest, although they were often working outside the forest reserves (Table 4.8).

In 1992 and 1993, a group of British biologists, under the Cambridge-Tanzania Rainforest Project surveyed six areas[93] of previously uninvestigated lowland forest in East Usambara. After presenting their results and through consultation with Kambai Village government, Muheza District government, TFCG, EUCADEP and EUCFP they were advised and welcomed to start the Kambai Forest Conservation Project (KFCP) under the auspices of TFCG. KFCP fell within the EUCADEP and EUCFP areas, but TFCG were welcomed since Kambai village is in a remote area, in which both projects had had difficulty in being effective. TFCG started work in Kambai village in 1994[94] with a member of EUCADEP,[95] Aidano Makange[96] seconded as Project Manager. Box 4.4 lists KFCPs major activities. TFCGs principle aim through KFCP, as with EUCADEP, was to contribute to the conservation of forest through its activities, which were geared to offering alternatives to forest products and forest land, and education. TFCG perceived its role to be guardian of the forest (Table 4.8).

With much of the forest on public land around Kambai having been gazetted as a reserve in 1992 with boundaries being extended in 1993, Kambai villagers were concerned that further areas of forest would be later included in the forest reserve. Several villagers had also had fields on the forest-field interface incorporated into the forest reserve. Compensation for the losses of the returns from labour were meant to be paid by the State,

---

support from the European Community (EC), and implementation support from the World Conservation Union (IUCN).

[93] Kambai and Semdoe forest were surveyed prior to their gazettment.

[94] I started as Project Co-ordinator in January 1995.

[95] EUCADEP already had a representative working in Kambai village, but he was not chosen by EUCADEP to be seconded to TFCG. TFCG initially tried to work in collaboration with this representative (Village Co-ordinator), but failed. This failure was due to the fact that on the few occasions that he decided to join us, villagers tended to ignore him. On one occasion an elder asked him: "Why are you here now? You have spent everyday of the last six years sitting outside your house, being paid for nothing. You only come now, because these people (TFCG) are showing you what you should have been doing. Go now, leave my farm!"

[96] In 1997, Makange's secondment ended. EUCADEP had been unable to pay their workers for some time and he became a full time member of TFCG. EUCADEP closed in 1997.

but to date many individuals have not been paid. This led to escalation in the clearing of forest on public land for agricultural land, before further areas of forest could be incorporated into the forest reserve and public rights to forest could be taken away.

Box 4.4 List of major KFCP activities

- Farm forestry;
- Improved land use management; and
- Environmental education.

Source: TFCG 1993-1998.

The village government unofficially allowed forest to be cleared by villagers. However, in 1996, Kambai villagers discovered that the EUCFP Forest Guard had paid for the illegal pit sawing of three ***mvule*** (*Milicia excelsa*) trees in the remaining ten hectares of Kambai public forest. Villagers informed TFCG staff, explaining that they perceived KFCP to be responsible for assisting the villagers in protecting forest and trees on public land. TFCG advised villagers to report these activities to the village government and in turn advised the village government to report this illegal pit sawing to the district and ward governments and EUCFP forest sub-station. This they did and the sub-station Forest Officer came to see his Forest Guard and ask if the allegations were correct. The Forest Guard denied them, but was warned that other Forest Officers would come to investigate.[97]

A couple of months later the EUCFP Manager, visited Kambai and held a meeting with the Village Chairman and Forest Guard. The Village Chairman told the Project Manager: "*If that man*[98] *stays to steal **our** trees,*[99] *I will instruct all villagers to cut down **your** trees.*[100]" The Forest Guard was forced to leave the village and was demoted to nursery worker. The statement of the Village Chairman indicates clearly the difference in relationship that villagers' felt they had with the public forest and the forest reserves: the public forest was **ours**, as in the villagers and the forest reserve was **yours,** as in the Forestry Divisions (Table 4.8). The villagers

[97] Several villagers felt that the Forest Officer and Guard were good friends and believed that the Officer warned his friend so that he could remove the incriminating timbers from his home before the investigation.

[98] 'That man' refers to the Forest Guard.

[99] 'Our trees' refers to trees on Kambai public land.

[100] 'Your trees' refers to trees in Kambai Forest Reserve.

were reaffirming their customary right to take on the responsibility of managing the public forests (Table 4.9).

Within Kambai public lands several individuals (Chapter Three) have retained natural forest of a few hectares for forest products, such as building poles, ropes, firewood and medicine (Table 4.8). In Magoroto Hill public lands there are areas of forest that have been retained under which to plant the cash crop cardamom (Table 4.8), since cardamom requires the shade of forest to grow effectively. As is customary, Makonde men continue to meet in the forest to conduct the initiation rites and circumcision of young boys. The Makonde in Kambai meet in secret in the public forests surrounding the village (Table 4.8).

Kwezitu public forest is between 100 and 200 hectares and is in two parts. Each part was customarily a clan forest. Members of Gonja, a sub-village of Kwezitu village, customarily manage one part. In 1997 members of Gonja visited TFCG staff in Kambai on three separate occasions and requested assistance in starting their own individual tree nurseries for tree planting on farms and along watercourses and hilltops (Table 4.8). They told TFCG that they had received free tree seedlings from EUCADEP in the past, but the quantity was insufficient. They had seen individual and group nurseries of their friends and family in Kambai and wanted their own nurseries in order to become self-sufficient. TFCG wrote to EUCADEP to inform them that TFCG staff would be visiting farmers who had requested advice. TFCG also asked to meet EUCADEP in order to discuss better collaboration of the projects.[101]

The week before TFCGs meeting with Gonja sub-villagers, George a Gonja sub-villager visited TFCG in Kambai. He told TFCG that EUCADEP project managers had arrived in the village and immediately called a meeting. He then relayed what had occurred in the meeting. Box 4.5 relays part of the dialogue from the meeting. It is ironic that this was the message that the villager gave the 'conservationist'. The forest and villagers had become belongings over which conservationists fought. But more importantly, the relationship between the local community and the forest had become that of guardians of the forest for ecological services (Table 4.8).

Shortly after this meeting, EUCADEP closed down due to internal problems and TFCG has since worked with Gonja sub-villagers in planning the management of their public forest (Table 4.9); managing individual tree

[101] Initially, collaboration between TFCG and EUCADEP had been good. Later collaboration became none existent when there was a change in EUCADEP management and all attempts by TFCG to arrange meetings failed.

Box 4.5 Dialogue from a meeting between EUCADEP and Gonja villagers, a sub-village of Kwezitu

EUCADEP:	"What is the name of that project down the hill?"
George:	"Kambai Forest Conservation Project."
EUCADEP:	"Yes. And where are we now?"
George:	"Kwezitu."
EUCADEP:	"Exactly, Kwezitu. So why do you want to work with them in Kambai?"
George:	"We are here in Kwezitu on the ridge above Kambai. If we cut all the forest of Kwezitu and all the trees along our streams, then it is not only us that will suffer, but our friends and family in Kambai, whose rivers rely on our streams. So that project is not only for Kambai, but for Kwezitu too."

Source: Author's fieldwork 1994-1998.

nurseries; and enrichment tree planting on hilltops and around springs and watercourses.

Forest on Private Land (Magrotto Estate, SHUWIMU)

Magrotto Estate stopped production in the 1990s. In 1991 (Tye 1993), the estate carried 269 hectares of mature oil palms, 35 hectares of young hybrid oil palms, some 20 hectares of cardamom cultivated under a thinned forest canopy with a cleared understorey and half a hectare of cloves. In addition, derris was formerly cultivated under the palms and still existed in large areas and there were small areas of black pepper that were grown on *Cedrela odorata* trees for support. The approximate area of the forest was between 200 and 300 hectares. The forests are however, heavily degraded by the planting of cardamom by local communities who have encroached on the estate forest (Table 4.8).

In the participatory era, SHUWIMU had still not utilised the land it leased although newspaper reports suggested that an orange plantation was planned (Nipashe 1996). Local communities continued to hold no rights to forest on SHUWIMU land (Table 4.9 & 4.10) and were classed as squatters (Table 4.8). This was despite the fact that the land had not been managed for over 20 years and local communities such as the Kambai sub-villages of Msakazi and Kweboha had been living and farming on the land for the same period (Table 4.9 & 4.10). However, the District Commissioner had advised Kambai villagers not to be concerned about their weak tenure and to plant permanent crops, such as trees, since he would back them in any quarrels over land tenure (Table 4.9).

Tree Tenure Regimes

One of the principle implementations of the participatory era was farm forestry. Both the State-run projects such as EUCFP and EUCADEP and projects of local NGOs such as TFCG implemented farm forestry in their project areas. Several Kambai villagers had already retained and planted trees in their fields and around their homes, but still welcomed advice and assistance from TFCG in tree growing (Chapter Three).

In Kambai tree tenure was weak where land tenure was weak. Women who had access to land via their male relatives had the right to plant and use trees in fields they held user rights to. However, the right to own, inherit or dispose of trees was held by the male clan member who had given them access to land. So the right of a woman to dispose of a tree she had planted often depended upon her relationship with her husband, father or brother. For instance, Mama Samora of Kweboha sub-village had the right to make all management decisions on the land she had been given by her husband. Mama Assia of Kambai who was a widow also had planted both a quarter of a hectare of teak (*Tectona grandis*) in one of her fields. Her land tenure was stronger than other women, because she as she explained, her husband and herself were originally from Mbeya region and had no other family living in Kambai. She explained that when he died she held the rights to the land because there were no male relatives to reclaim their clan land. In this way, her rights to trees were also strong.

In the sub-villages of Kweboha and Msakazi that were on SHUWIMU land, land tenure was weak and hence some villagers were apprehensive about planting trees. They were concerned that they may not see returns from their labour if they were to be moved from the land. However, with the back up of the District Commissioner who came to the village to talk about such issues, many villagers felt safe enough to plant perennial crops such as trees.

Although in general, tree tenure was highly correlated to land tenure in this area, there were exceptions. The rights of government protected trees, such as ***mvule*** (*Millicia excelsa*) were held only by the State and even if a tree had been planted by a villager in his or her field then the right to dispose of the tree laid with the State. A permit was required for its disposal. One old man even admitted to pulling up ***mvule*** (*Millicia excelsa*) wildlings in his fields. He explained that he did not want such a tree in his field, even though it was good for timber, if he did not have the right to obtain timber from it without first obtaining an expensive permit from the

State. Several Kambai villagers decided to plant ***mvule*** (*Millicia excelsa*), with seedlings from TFCG. They planted in a straight line in the hope that it would be obvious that they had planted them and hoped that in the future the policy would be changed and that they would have the right to dispose of trees which they themselves had planted.

Summary of Stakeholder Relationships, Rights and Responsibilities to Forest in Different Management Eras

Stakeholders' relationships, and rights and responsibilities to the forests of East Usambara and Udzungwa have changed through different forest management eras and are summarised in Tables 4.11 and 4.12. It is interesting to note that through time the number of categories of stakeholders has increased (Tables 4.11 & 4.12). For instance, in the local customary era, the only category of stakeholder was the local community[102] (Tables 4.11 & 4.12). In the technocratic era categories of stakeholders included local communities, the State, and the Private Sector and International Community (Tables 4.11 & 4.12). In the participatory era categories of stakeholders included all those in the technocratic era, plus NGOs such as TFCG (Tables 4.11 & 4.12).

It is also important to note that in both the technocratic and participatory eras, customary and statutory roles coexisted (Tables 4.11 & 4.12). Several of these customary roles were derived from the local customary era, but since this Chapter demonstrates customary roles to be dynamic, they also evolved in the technocratic and participatory eras (Tables 4.11 & 4.12).

Although stakeholders' roles have been analysed via natural forest management eras, it is important to note that management approaches have often coexisted in parallel. For example, in the technocratic era, local customary management approaches were demonstrated to continue well into the technocratic era and in the participatory era, further reservations of forest - a typical intervention of the technocratic era - continued.

Local Customary Era (1740-1892)

In the local customary era, the relationship between local communities and forest was ambivalent and the forest was therefore both respected and feared (Table 4.11). Forests were places believed to have the power to heal

[102] The local community is not homogenous (Chapter Two).

Table 4.11 Summary of stakeholder relationships[103] to forest in different management eras[104]

	Natural Forest Management Era		
Stakeholder	Local customary (1740 - 1892)	Technocratic (1892 - 1989)	Participatory (1989 - 1998)
Local community[105]	[C] Ambivalent relationship to forest. The forest was both feared and respected. [C] Philosophy of conservation, whereby ritual sacrifice was required to heal the land.	[S] Relationship to forest reserves and private forests broken. [S] Squatters and thieves of reserved and private forests. [S] Beneficiaries of forest on public land. [C] Initially local customary relationship maintained. [C2] Immigrants bring their own customary relationships to forest. [C2] In 1940s and 1950s local customary philosophy of conservation eroded. Relationship that of beneficiary of forest resources only.	[S] Relationship to forest reserves and private forests broken. [S] Squatters and thieves of reserved and private forests. [S] Beneficiaries of forest on public land. [C2] Immigrants maintain their own customary relationships to forest. [C3] Forest reserves perceived purely as domain of State. [C3] Beginnings of guardianship of forest on public land and LGFR. [C] [C2] [C3] Several customary benefits of forest maintained and negotiated.

103 Where [C] denotes customary relationship evolved in local customary era; [C2] denotes customary relationship evolved in technocratic era; [C3] denotes customary relationship evolved in participatory era; and [S] denotes statutory relationship.

104 See Chapter Two, Table 2.3 for framework of analysis: natural forest management eras and dates.

105 The term local community is consistent with the term village, where village refers to the social community and the area of land belonging to the community.

State[106]		[S] Forest resources primarily valued for commercial worth. [S] Forest valued for ecological services. [S] Guardians of forest.	[S] Forest valued primarily for biodiversity and ecological services. [S] 'National trees' reserved by State for the potential value of timber returns. [S] Guardians of forest.
Private Sector & International Community[107]		[S] Forest resources primarily valued for commercial worth.	[S] Forest valued primarily for biodiversity and ecological services.
TFCG			[S] Forest valued primarily for biodiversity and ecological services. [S] Guardians of forest.

Source: Author's fieldwork 1994-1998.

or harm. This power was linked specifically to the belief that forest trees were the abode of ancestral spirits and it was the ancestral spirits whose connection with God held the power to heal or harm. In order to ensure that the ancestors worked on the side of healing rather than harm, local communities were required to ensure the ancestors did not become angry by carrying out ritual sacrifice. Paying tribute to leaders,[108] who customarily carried out the rituals, ensured ritual sacrifice and ancestors benevolence towards local communities, therefore bringing rain rather than drought and healing the ill rather than letting them die.

Local communities have been demonstrated to have a philosophy of conservation (Table 4.11), whereby ritual sacrifice was required in order to heal the land or preserve and protect the environment. Ritual forests were

[106] The State has ranged from the colonial administrations of the Germans and British to be an Independent State.
[107] The private sector includes estates & the international community includes researchers & donor funders.
[108] Prior to Kilindi political control (pre 1740) it was not necessarily the leaders who held rain charms.

Table 4.12 Summary of stakeholder rights and responsibilities[109] to forest in different management eras[110]

	Natural Forest Management Era		
Stakeholder	Local customary (1740-1892)	Technocratic (1892-1989)	Participatory (1989-1998)
Local community [111]	[C] Access and user rights to forest resources on the condition that forest rules are obeyed. [C] Responsibility of leaders to make and uphold forest rules and responsibility of community members to adhere to rules.	[S] No rights to forest reserves and forest on private land. Responsibility to adhere to rules. [S] Access and user rights to forest products in forest on public land. No responsibilities. [S] Right to pit saw with permits from State in forest on public land. [C] Prior to the 1940s and 1950s rights and responsibilities as in the local customary era. [C2] In 1940s and 1950s local customary rights and responsibilities eroded. Right to access and use forest, without corresponding responsibilities. [C2] [C3] Right to access and use	[S] No rights to forest reserves and forest on private land. Responsibility to adhere to rules. [S] Access and user rights to forest products in forest on public land. No responsibilities. [S] Right to pit saw with permits from State in forest on public land. [C2] [C3] Responsibilities eroded. Access and user rights to forest products, without corresponding responsibilities. [C3] Local communities starting to take more responsibility for the management of forests, with moral support from TFCG. [C2] [C3] Right to access and use

109 Where [C] denotes customary roles evolved in local customary era; [C2] denotes customary roles evolved in technocratic era; [C3] denotes customary roles evolved in participatory era; and [S] denotes statutory roles.

110 See Chapter Two, Table 2.3 for framework of analysis: natural forest management eras and dates.

111 The term local community is consistent with the term village, where village refers to the social community and the area of land belonging to the community.

	forest on estate land that has been unused for a long period of time.	forest on estate land that has been unused for a long period of time.
State[112]	[S] Right and responsibility to control forest reserve access and use. [S] Responsibility to employ staff to guard forest reserves from illegal access and use. [S] Right and responsibility to permit or prevent the felling of government protected trees. [S] Right and responsibility to demand forest management plans for forest protected on private lands.	[S] Right and responsibility to control forest reserve access and use. [S] Responsibility to employ staff to guard forest reserves from illegal access and use. [S] Right and responsibility to permit or prevent the felling of government protected trees. [S] Pitsawing banned in forest reserves.
Private Sector and International Community[113]	[S] Rights to and responsibility to obtain logging licences.	[S] Right to undertake research in the forest reserves with research permits. [S] Right and responsibility to manage forests protected on private lands.

[112] The State has ranged from the colonial administrations of the Germans and British to be an Independent State.

[113] The private sector includes estates and sawmills and the international community includes researchers and donor funders.

TFCG[114]		[S] Right to undertake research in the forests with research permits. [S] District council appointed TFCG as Forest Manager of Lulanda LGFR with the right and responsibility To manage the forest on behalf of the State.

Source: Author's fieldwork 1994-1998.

selectively reserved for ritual sacrifice, whereas clan forests were often managed more for local returns. The local community would not however, fell a tree, clear forest or collect forest products without first seeking to 'cool the anger of the spirits' through ritual sacrifice. That it was the leaders, who held ritual power and managed the conservation of the forests, demonstrates that the relationship between the local community and the forest was based not only on the belief system but also on the political system.

Forest tenure regimes were hierarchical and were secured through social relationships. In Usambara for example, all land was 'owned' in its political sense by the Kilindi king. For the Sambaa, ownership was thought to ensure the benevolent use of ritual power (Feierman 1974). Hence, the king was given the wealth of all the land, through tribute, so that ritualsacrifice was ensured, as was a benevolent relationship with ancestral spirits who would therefore make rain. Since the ancestral spirits were thought to abide in specific forests – ritual forests – these forests were controlled by Kilindi clan leaders who held the special rain charms. Since leaders from each clan in a community were required to carry out ritual sacrifice, these clan leaders too had authority in making and upholding forest rules (Table 4.12).

In case studies of East Usambara ritual forests access and user rights were also given to community members and it was the responsibility of community leaders to uphold and punish those not adhering to forest rules (Table 4.12). In contrast, in the case study of Lulanda in Udzungwa, access

[114] The Tanzania Forest Conservation Group (TFCG) is a local NGO.

to ritual forest was for leaders only. In the case of clan forests, clan leaders held authority and tenure was secured by being and remaining a member of the clan. Women secured their tenure through their relationships with men, for example, as daughters and wives. However, upon leaving the clan through marriage or divorce, they would no longer be a member of that clan and would have to secure tenure with their husbands' clan or return to their fathers' clan.

Technocratic Era (1892-1989)

In the technocratic era, the State, private sector and international community valued forest resources primarily for their commercial worth (Table 4.11). The State, private sector and international community associated tenure security with its spatial aspects. It was for these reasons that State policies reserved and protected forests, although these policies were often backed by ecological and biological arguments (Table 4.11). The forest tenure regime was hierarchical in that the State had ultimate control over rights to access and use the forests even where forests were on public land and especially where timber tree species were concerned (Table 4.12).

State perceptions of local community relationships to the forests were also convenient in backing reservation policies. Initially, the State mistakenly believed that local communities had no relationship with some forest areas, purely on the basis of differences in African and Western tenure regimes and that any relationship they did have with forest was only as beneficiaries of forest resources. They therefore reserved forest safe in the belief that "*...forest resources are far in excess of the local demand, and that this probably always will be the situation*" (Grant 1924). When clashes arose between the State and the local community, the opposing argument was used to justify further reservations "*...African forest use, which threatens the entire colony's timber supplies, is unjustifiable*" (Troup 1936). The State's perception of local communities' relationship with forest as purely beneficiary was however taken into consideration and the forest on public land or "*...remaining forest was for the benefit of the local people...*" (Legislative Council 1953).

With the reservation of forests, local communities' customary relationships were officially broken (Table 4.11). Those communities maintaining customary relations in the reserves and in private forests were officially squatters and thieves of forest resources (Table 4.11) and ironically were often assisted in these illegal practices by forestry staff after offering bribes. Local communities' relationships became instead of managers of forests, beneficiaries only. They then often sought to gain returns from forest resources at an accelerated rate before favourable conditions changed and/or forestry staff moved on.

In the forests on public land customary relationships were initially maintained as in the local customary era (Table 4.11) and immigrant estate workers also brought with them their own unique customary relationships to the forest (Table 4.11). A dualistic tenure regime existed with both statutory and customary tenure regimes coexisting. Local customary relationships and tenure regimes were demonstrated to exist up until the 1940s and 1950s, but elders remember that it was around this time that the authority of customary leaders was eroded (Tables 4.11 & 4.12). They had no power and hence no control over the management of the forests. In this way the forests, which had been customarily closed, with community leaders controlling access and use, had become open, with community leaders being powerless to control forest access and use (Table 4.12). Local communities' relationship with forest had become that of purely beneficiaries and the perceived understanding of local-forest relationships by the State had been fulfilled and hence perpetuated.

Participatory Era (1989-1998)

In the participatory era, high biodiversity and ecological values of forest were increasingly recognised by the international community (Table 4.11), catalysed by awareness of extensive logging in East Usambara in the technocratic era. The international community increasingly put pressure on the State, through donor funding, to conserve forest for these values (Table 4.11). Due to increased awareness and donor funding in conservation of forest, several conservation and environmental NGOs began to focus on the Eastern Arc forests, such as the World Conservation Union (IUCN)[115] and TFCG (Table 4.11).

The State and international community continued to view tenure security in respect to its spatial aspects. Therefore with increased focus on the forests for their high biodiversity value, forests continued to be gazetted as reserves, following the dominant management approach of the technocratic era. The forest tenure regime remained hierarchical with the State controlling rights and responsibilities to access and use forest and government protected timber trees (Table 4.12). In the meantime, forests continued to be customarily open as in the technocratic era, with local

[115] IUCN funded and gave technical assistance to EUCADEP.

communities accessing and using forests without any corresponding responsibilities (Table 4.12). Local communities' relationships with the forest continued to be broken and were often officially squatters and illegal users of forest resources (Table 4.11). The result of increased reservation of forests was further encroachment and degradation of forest, as local communities sought to gain returns from forest, before further forest was reserved (Table 4.11).

This perpetuated the perception that the only relationship between local communities and forest was that of beneficiary of forest resources (Table 4.11). Local communities were therefore perceived as the problem to achieving sustainable forest management. The solution was to implement community-based projects with key interventions that offered alternatives to forest products, such as through farm forestry; reduced pressure for forest land by improving land use management; and educated locals as to the ecological benefits of conserving forests, through environmental education.

Although local communities welcomed these interventions in terms of sustainable development, in general, the interventions themselves did not achieve the aim they were created for, principally that of forest conservation. For instance, a Kambai villager who had planted over 400 tree seedlings in his fields proceeded to clear two hectares of public forest for further agricultural land. Stating that he had done so, before the State could take it as forest reserve. Another, Kambai villager felt that over half of the local community understood the benefits of conserving forest, but would still clear and degrade forest as long as it meant lost potential had been regained.

In several cases where the State's efficiency in forest conservation was reduced due to lack of funds, NGOs, such as TFCG, took on the role of assisting the State in the guardianship of forest reserves, for instance, Lulanda LGFR (Table 4.11). With increased biological research in many of the forests, the failure of the State to manage and protect many of the forests was realised. The District authorities that controlled Lulanda LGFR accepted the assistance of TFCG a local NGO, in managing the forest. TFCG were responsible for managing the forest, but ultimate authority remained with the District (Table 4.12). The TFCG initially managed the forest along technocratic lines following the approach of the State of guard and protect.

However, because TFCG were community-based and held meetings with the Lulanda community to discuss forest issues, power was increasingly localised even though it was not the community itself which held the power. Through these discussions the Lulanda community took on the responsibility of assisting TFCG in the work of management, even without corresponding rights. TFCG however, informally allowed the customary collection of medicinal plants and **ulanzi** (bamboo alcohol) (Table 4.12). However, several Lulanda villagers still remained bitter about their lack of power. Stating that the forest was "*no longer their concern*".

In the latter part of the participatory era, local communities described themselves as 'thieves of the forest reserves' since they had no official relationship (Table 4.11). Through a series of discussions between TFCG and the Kambai community, the beginnings of a guardianship relationship to forest on public land appeared to develop (Table 4.11). This was demonstrated when the Kambai community reported the forest reserve guard for pitsawing in what they called 'their public forest'. TFCG also gained the assistance of the Lulanda community in managing Lulanda LGFR. Through discussion, Lulanda elders requested that TFCG allow the customary collection of medicinal herbs and **ulanzi** (bamboo alcohol) and the use of customary pathways through the forest (Table 4.11). These agreements were not formalised, but were the beginnings of the negotiation of forest management between stakeholders. Gonja sub-villagers were also reviving customary control and management of part of Kwezitu public forest (Table 4.12).

Summary

It is ironic that through time, the stakeholders who are physically closest to the forest - the local community - have become the stakeholders whose official relationship with much of the forest is the most distant. It is also ironic that the demonstrated local philosophy of forest conservation was eroded in colonialism through colonial politics and religion, to be replaced by local communities with misconstrued Western relationships to forest. It is also clear that interventions that were developed on the assumption that local communities' were purely beneficiaries of forest resources did not succeed in sustainably managing forest.

Tenure regimes that had been socially defined in the local customary era had been spatially and economically defined in the technocratic and participatory eras. Power to control forest rights and responsibilities had moved from local community-based authority in the local customary era to District and Central government authority in the technocratic and participatory eras. In the participatory era, the increasing responsibility of the State put on by the International Community to manage the forests sustainably led to the State seeking to share management responsibilities with NGOs, such as TFCG. The local NGOs that were community-based themselves sought the co-operation of the local communities, either by trading access and user rights for the responsibility of the work of management. Neither the rights nor responsibilities to manage or control the forests were being traded. However, what NGOs like TFCG which were locally community-based did do, was to give communities the opportunity to openly discuss issues concerning the management of the forests and obtain impartial advice and assistance, when and if required.

Chapter 5

Conclusion: Implications for the Development of Sustainable Forest Management Practices

Introduction

For a decade or more there has been increasing awareness that more effective approaches for managing natural forests in the developing world are required, and recognition that such approaches almost certainly need to involve forest-local communities (Chapter One). There is however less agreement as to what role forest-local communities should play in the management of the forests and particularly those of high biodiversity, such as the Eastern Arc.

Previous research into the role of forest-local communities in natural forest management in Tanzania has concentrated on their returns from forest resources, since their role has been seen as purely beneficiaries of forest resources (Chapter One). More recently, over the last three or four years, practical work in the field by State Forestry staff and NGO staff in particular, has begun to focus more fully on the role of forest-local communities in natural forest management. So far, roles have often been delimited to user rights and the responsibility of the work of management (Chapter One). There are however, several cases of community forest management emerging, but these have been in the Miombo woodlands or in forests on public or private land. All forests which are not valued particularly for their high biodiversity.

This research has therefore examined changes in stakeholders' roles through different management eras[116] in the high biodiversity forests of the Eastern Arc Mountains. Roles were not only analysed in respect to returns

[116] See Table 1.1 (Chapter One) which contrasts different natural forest management approaches and is central to conceptualising change in forest-local communities' roles in forest management.

from forest resources (Chapter Three), but also via stakeholders' rights, responsibilities and relationship to forest (Chapter Four), and the consequential effect on the sustainable management of forests.

This research contributes to an understanding of the effect of imbalances in stakeholders' roles on the development of sustainable forest management practices historically. The findings of this research offer implications for a constructive negotiation process, just as the 1998 Tanzanian National Forest Policy is put into practice and stakeholders' roles in the management of the Eastern Arc forests must begin to be negotiated.

Effect of Imbalances in Stakeholders' Roles on Sustainable Forest Management

Natural forest management approaches in Tanzania have evolved through time. Local customary, technocratic, participatory and political negotiation approaches to forest management have been identified in the Eastern Arc (Chapter Two). For the purpose of analysis, these approaches have been placed historically in eras (local customary, technocratic, participatory, and political negotiation), where an era is identified by the predominant forest management paradigm of the time (Chapter Two: Table 2.3). It is important to note, however, that forest management approaches were often coexisting within one period identified as an era. This shall be discussed more fully here by examining how coexisting approaches rebounded on each other. The research has analysed changes through the local customary, technocratic and participatory eras only, because the Tanzanian National Forest Policy - enabling negotiation - was only put in place in 1998 at the end of the research period. The implications of the findings for a constructive negotiation process, are however, discussed in the following section.

Table 5.1 summarises the findings of the research in relation to the local customary, technocratic and participatory forest management eras, highlighting imbalances in and between stakeholders' roles; forest management authority; the definition of forest used by management authority; and the sustainability of forest management.

The first thing to note is that the number of major stakeholders in Eastern Arc forest has increased through changing management eras (Table 5.1). In the local customary era, the major stakeholder was the forest-local

Table 5.1 Summary of research findings

	Natural Forest Management Era		
	Local customary	Technocratic	Participatory
Forest-local community balance of roles[117]	[C] Relationship [C] Rights (Management) [C] Responsibilities [C] Returns	[S] Returns [S] Rights (User and Access) [C][C2] Returns	[S] Returns [S] Rights (User and Access) [C2][C3] Returns [C3] Responsibilities [C3] Relationship
State balance of roles		[S] Relationship [S] Rights (Management) [S] Responsibilities [S] Returns	[S] Relationship [S] Rights (Management) [S] Responsibilities [S] Returns
Private Sector/ International Community balance of roles		[S] Relationship [S] Rights (Management) [S] Responsibilities [S] Returns	[S] Relationship [S] Rights (Management) [S] Responsibilities [S] Returns
TFCG balance of roles			[S] Relationship [S] Rights (Management) [S] Responsibilities
Forest management authority	Forest-local community leaders	State	State
Forest management defined by authority	Culturally Socially Economically Politically	Economically	Economically Ecologically Biologically
Sustainability of forest management	Locally managed.	Managed and unmanaged deforestation and forest degradation.	Unmanaged deforestation and forest degradation.

Source: Author's fieldwork 1994-1998.

community. In the technocratic era major stakeholders included the forest-local community, State and Private Sector and in the participatory era

[117] [C] represents customary roles developed in local customary era; [C2] represents customary roles developed in technocratic era; [C3] represents customary roles developed in participatory era; [S] represents statutory roles.

major stakeholders included all those in the technocratic era, plus NGOs and the International Community. This demonstrates a move from, locally managed forests in the local customary era, to an increasingly centralised, distanced management of forests. As Chapter Four demonstrates the effect on the management of the forests has been negative, with forests that were previously managed locally becoming increasingly deforested and degraded (Table 5.1) despite State policies and management approaches designed to conserve forests. The crux of the matter is that through changing forest management eras stakeholders' roles have become unbalanced (Table 5.1) and the relationships between stakeholders have consequently become strained.

The Eastern Arc forests are significant globally, nationally and locally, but their significance and to who has varied through changing management eras. This has been demonstrated to be a major constraint to the effectiveness of management approaches through the eras (Chapter Four). Forest that was defined by management authorities, namely local community leaders, politically, socially, economically and culturally in the local customary era, were defined by the State, predominantly politically and economically in the technocratic era, and politically, economically, ecologically and biologically in the participatory era (Table 5.1). Management approaches developed on one stakeholders' definition of forest alone have been demonstrated to lead to the coexistence of management practices, which may cause further confusion and mistrust between stakeholders. A situation, that may ultimately lead to the development of further unsatisfactory forest management practices.

This is demonstrated by the consistent belief of the State that forest-local communities' definition of forest is economic only and consequently their role in forest management is purely that of beneficiaries (Chapter Four). Hence, the forest reservations of the technocratic era and farm forestry interventions in the participatory era. The State's preconception of forest-local communities' roles dates from the technocratic era. However, the State has perpetuated its own perception by the intervention of reservation. State authority of forest management, with the removal of local customary management, access and user rights and responsibilities, led to the erosion of local customary management practices (Chapter Four). New customary rights were developed however, as forest-local communities sought to gain returns based on misconceptions of Western tenure regimes (Chapter Four). However, with the erosion of local customary management authority, these new customary rights often had no corresponding responsibilities (Table 5.1). Subsequently forest tenure has moved from closed to open access in real terms and led to deforestation and forest degradation (Chapter Four).

In the participatory era, often the technocratic intervention of reservation remained predominant, combined with farm forestry interventions. The focus being again, forest-local communities' as beneficiaries of forest resources. That these approaches failed can be highlighted by the example of a forest-local community member planting over 400 tree seedlings on his farm and then clearing a large area of public forest (Chapter Three).

In some of the more progressive management practices of the participatory era, forest-local communities have been offered access and user rights to minor forest products in return for labour responsibilities in managing forests. For example, see the case study of Lulanda LGFR (Chapter Four). However, in offering access and user rights to forest resources, the focus of forest-local communities' roles was again on returns from forest resources (Table 5.1). As Chapter Three demonstrates, although forest-local communities sometimes prefer to use forest resources for certain products (for example, building poles) they do not depend on forest resources. Forest-local communities' admit to allowing forest degradation by obtaining returns in areas for which they have no management authority or responsibilities. In this way they once again gain control over forest resources, for which they would otherwise have no control.

What has been achieved in the participatory era however, is greater discussion and contact between stakeholders, both State, NGOs and forest-local communities. This has led in some cases to forest-local communities gaining greater confidence in beginning to develop new customary management practices in the Eastern Arc forests. This was seen particularly in the public forests (for example, see the case study of Kwezitu) and when in conjunction with the TFCG and Mufindi District Government in Local Government Forest Reserves (for example, see the case study of Lulanda LGFR).

If the management of Eastern Arc forests is to be more sustainable then the role of forest-local communities cannot be based on access and user rights and the responsibility of labour in managing forest alone, but must move towards a more meaningful role of authority, control and responsibility. Stakeholders in forest management need to be clearly identified. Identification of both primary and secondary stakeholders is required and management authority given to those stakeholders who are best able to manage the forests practically. Forest must be defined holistically: politically, socially, economically, culturally and ecologically and biologically. Sustainable forest management has to be defined not as the biomass resource per se, not as economic production of the forest, but as a new political and social contract that guarantees human rights necessary for survival and dignified living. This research highlights the need for local political negotiation to ensure sustainability in forest management.

Implications for the Development of Sustainable Forest Management Practices

As the findings have demonstrated stakeholders in forest management need to be clearly identified and their roles negotiated if sustainable forest practices are to be developed. The new Tanzanian National Forest Policy, announced in 1998, enables negotiations between stakeholders through the development of both Joint Forest Management and Community Forest Management agreements (Chapter One). This section discusses the implications of the findings for the development of a constructive negotiation process for forest management.

To achieve a constructive negotiation process, capacity needs are more institutional than technical. They can be divided into two categories: the capacity for negotiation itself; and at a later stage, capacities for sustaining roles. The capacity for negotiation can depend on the:

- Willingness of stakeholders to accept the negotiation process as a means to more sustainable forest management;
- Willingness of State, NGOs and donor funders to put resources such as time, money and expertise into the process of negotiation;
- Availability of competent and experienced independent facilitators of negotiation;
- Capacity for stakeholders to negotiate, by empowering the weakest stakeholders, particularly through the provision of information; and
- Realisation that negotiation itself must be seen as an ongoing learning process, as stakeholders develop their roles through time and changing circumstances.

Once the process of negotiation has begun, further practical implications come to the fore. The capacities for sustaining roles can depend on the:

- Accountability and representativeness of local governance and leadership; and
- Local development of progressive laws that support more sustainable management practices.

The willingness of stakeholders to accept the negotiation process as a means to more sustainable forest management is crucial to the development of a constructive negotiation process. As the findings have demonstrated, if stakeholders continue to misunderstand and mistrust each other, predefining other stakeholders' definition of forest and role in forest management, then predefined problems and solutions are often all that are negotiated. For instance, in the participatory era, access and user rights, and the responsibilities of the labour of management were negotiated, but nothing more. Both the State and the NGO predefined the limits of what was to be negotiated. The 1998 Tanzanian National Forest Policy also seems to predefine the limits of negotiation. The policy moves towards more meaningful roles for forest-local communities, although the extent of their roles depends upon the official status of the forest in question. What is offered in both Central and Local Government Forest Reserves are Joint Forest Management agreements, whereas in the forests on public and private lands Community Forest Management is a possibility. Since it is usually the reserves which are valued most for their high biodiversity, what this policy seems to suggest is that forest-local communities cannot be fully trusted to manage forest reserves alone. Hence, the negotiation process itself does not seem to be perceived by the State to be a means in itself to more sustainable forest management.

The willingness of the State, NGOs and donor funders to put resources such as time, money and expertise into the process of negotiation relates partially to the willingness of these stakeholders to accept the negotiation process. NGOs such as TFCG have a mandate to conserve forests of high biodiversity in the Eastern Arc. They, like government institutions and donor funders, therefore may be reluctant to put time, money and expertise into the negotiation process, particularly in public and private forests that are perhaps not valued for their high biodiversity. They may however, be unwilling or wary about accepting that forest-local communities, amongst others, can play a meaningful role in the management of forests through JFM. Consequently, they may be slow to attempt to facilitate the negotiation process in forests of high biodiversity, preferring to wait to 'see the results' of CFM in less high biodiversity forests. However, without resources, such as time, money and expertise, the success of the negotiation process, whether in JFM or CFM, is likely to be much reduced. Also donor funders often work on three-year funding phases and are more interested in products rather than processes. This time and money constraint often puts pressure on facilitating institutions to come up with results in terms of, for example, 'number of trees planted'. Often these institutions take the 'easy option' and tend to concentrate on old familiar management practices, guarding the forest or tree planting, for example, expecting the negotiation process to develop on its own, with little input.

The availability of competent and experienced independent facilitators of negotiation could seriously jeopardise the negotiation process. The majority of State and NGO staff have been trained in technocratic approaches to forest management and are therefore unlikely to be competent and experienced facilitators. The history of mistrust and imbalance in power between certain stakeholders, demonstrated in the findings, particularly between the State and forest-local communities, requires that facilitators be independent.

The capacity for stakeholders to negotiate, by empowering the weakest stakeholders, particularly through the provision of information, is crucial for a meaningful negotiation process. The findings have demonstrated that often information is not clearly or correctly given to all stakeholders, particularly where the rights of forest-local communities are concerned. This has been demonstrated throughout the research, by forest-local communities and in fact the District Government, also, not being clear whether a forest such as Lulanda is actually a gazetted forest reserve or not. All stakeholders need clear, concise information, in Kiswahili, concerning their rights in forest management and the possibilities for the negotiation of their roles, brought about by the 1998 Tanzanian National Forest Policy. Newsletters, such as the Arc Journal, which are published and distributed to forest-local communities in Kiswahili, are useful starting points. However, the greatest provision of information is through community meetings and discussions.

It is important to realise that negotiation itself must be seen as an ongoing learning process, as stakeholders develop their roles through time and changing circumstances. Those who expect too much too soon may become frustrated. Those who are unwilling to accept the negotiation process as a means to sustainable forest management, may use difficulties in the negotiation process as an excuse to disregard it. However, through time, more and more positive case studies of forest management through the negotiation of roles should evolve.

The accountability and representativeness of local governance and leadership is important for sustaining roles. Although this research has focused on major stakeholder groups, such as forest-local communities, State and NGOs, it is also important to analyse imbalances in roles at the local level of forest-local communities. Participatory stakeholder analysis and the analysis of stakeholders' roles at forest-local community level, via their rights, responsibilities, returns and relationships is an important prerequisite of the negotiation process. Only then can accountability and representativeness of local governance and leadership develop. As forest management is moving from State to community management, perhaps community forest management will move from village leaders to ordinary villagers.

The local development of progressive laws that support more sustainable management practices are required for the sustenance of roles. Without laws to back up stakeholders' rights and responsibilities, there is likely to be a situation where dualistic roles, confusion and mismanagement of the forests continue.

Tanzanian forest management approaches are in the process of change, as are forest management approaches around the world. What is required is the development of a constructive negotiation process that will in turn contribute to the development of more sustainable management practices.

Appendices

Appendix 1: Medicinal Remedies used by Traditional Herbalists on Magoroto Hill

Illness/ Problem	Vernacular Name	Botanical Name	Type	Part	Source
Appendicitis	***Mtamba kuzimu***	*Deinbollia borbonica*	Tree	Roots	Shamba
(Roots boiled and water drunk in morning before food and drink.)					
Asthma	**Ufyambo**	?	Climber	Leaves	Forest
(Sap squeezed out of a handful of leaves, mixed with water, drunk.)					
Asthma in children	***Mhasha***	*Veronica iodocalyx*	Bush	Leaves	Shamba
	Muuka	*Microglossa densiflora*	Shrub	Leaves	Shamba
(Sap mixed with water and drunk. A spoonful taken morning, afternoon and evening for two days.)					
Earache	***Mchunga***	*Sonchus luxurians*	Herb	Leaves	Shamba
(Sap administered to ear in the evenings for two days.)					
Epilepsy	**Kamachuma**	*Priva cordifolia*	Shrub	Roots	Shamba
	Mlenga	*Veronia abconica*	Shrub	Roots	Shamba
(One root of each taken and boiled and infusion drunk once a day for one week.)					
Epilepsy	***Mti wa kondo***	?	Bush	Leaves & Roots	Shamba
	Hozandoghoi	*Hyptis pectinata*	Shrub	Leaves & Roots	Shamba
	Muuka	*Microlossia densiflora*	Shrub	Leaves & Roots	Shamba
	Mwinika nguu	*Asparagus falcatus*	Bush	Leaves & Roots	Forest
	Mlenga	*Veronica*	Shrub	Leaves	Shamba

		abconia		& Roots	
(Sap of leaves drunk with water and roots boiled and water drunk in the morning.)					
Eczema in Children	**Gugufa**	?	Shrub	Leaves & Roots	Forest
	Samaka	*Aframomium sp.*	Shrub	Leaves & Roots	Forest
	Ngoko	*Piper capensis*	Shrub	Leaves & Roots	Forest
	Ufiha	*Olyra latifolia*	Shrub	Leaves & Roots	Forest
	Ushuki	?	Grass	Leaves	Shamba
	Umpolo	*Draecaena laxissima*	Shrub	Leaves & Roots	Forest
	Kihumpu	*Mucuna pruriens*	Climber	Leaves & Roots	Shamba
	Mpingo	*Dahlbergia melanoxylon*	Tree	Leaves & Roots	Forest
(Leaves of all arc dricd, burncd and the ash is mixed with oil and applied to the skin morning and night for seven days. The roots are boiled and the water drunk morning and night also.)					
Eye ache	***Ndiga***	?	Shrub	Roots	Shamba
	Ghole	*Adenia cissampe-loides*	Climber	Roots	Forest
(Roots burned and ash added to a little water and stored for a day. Eyes are then washed with the solution once before bedtime.)					
Female Infertility	***Muungu***	*Saba florida*	Tree	Roots	Forest
	Kitengwazi	?	Tree	Roots	Forest
(Roots boiled and water drunk, morning, afternoon and evening for seven days.)					
Flatulence	***Mtamba kuzimo***	*Deinbolua borbonica*	Tree	Roots	Shamba
(Roots boiled and water drunk before breakfast. Only one dosage taken.)					
Headache	**Mshuza**	*Citrus aurantium*	Tree	Leaves & Roots	Shamba
	Hashaanda	*Veronia*	Shrub	Leaves	Shamba

		corolata		& Roots	
	Mdongonyezi	*Toddaua asiatica*	Tree	Leaves & Roots	Forest
	Mkuungo	*Terminalia sambesiaca*	Tree	Leaves & Roots	Forest
(Roots boiled and water drunk, morning, afternoon and evening for seven days. Leaves dried, ground to powder and applied to head once.)					
Impotency	***Mkweme***	*Telfairia pedata*	Climber	Fruit	Forest
	Kijamitu	?	Tree	Root	Forest
(Roots boiled and eaten with juice of Mkweme fruit.)					
Insanity	**Mwae**	?	Tree	Leaves	Forest
	Mvule	*Milicia excelsa*	Tree	Leaves	Forest
(Leaves of Mvule must have fallen naturally. Leaves are boiled and water is drunk three times a day for seven days. If the patient starts to get better, the patient is taken under a big tree, hair shaved off and dried burnt leaves of **Mwae** and ***Mvule*** are rubbed into the scalp and chest.)					
Intestinal Worms	***Muungu***	*Cola usambaren-sis*	Tree	Bark	Lowland Forest
	Mhombo	?	Tree	Bark	Shamba
	Mviu	*Vangueria tomencosa*	Tree	Bark & Roots	Forest
(Boil bark of each with roots of ***Mviu*** and drink water.)					
Invincibility	***Mjolwe***	?	Tree	Roots	Forest
	Mwembe	*Mangifera indica*	Tree	Bark	Shamba
	(Naturally fallen coconut tree)	?	Tree	Bark	Shamba
	Ngukia	(Type of stone)			
(Mjolwe root dried and powdered, mixed with dried powdered bark of Mwembe and coconut tree and then added to white powder from ***Ngukia*** stone. Take mixed in porridge in the evening for three days. If a person is killed by a snake bite or gun shot this remedy will resuscitate.)					
Invisibility	**Kingoza**	?	Tree	Roots	Shamba
(To hide from attackers, place root in mouth and close eyes.)					
Menstral pain	***Muungu***	*Cola usambaren-sis*	Tree	Bark	Lowland Forest
	Mhombo	?	Tree	Bark	Lowland

					Forest
	Mviu	*Vangueria tomencosa*	Tree	Roots	Forest
(Bark of ***Muungu*** and ***Mhombo*** boiled with a little water, roots of ***Mviu*** burnt. Ash mixed with bark infusion and drunk once a day for five days.)					
Mumps	**Mkuungo**	?	Tree	Bark	Lowland Forest
(Boil bark and drink water, morning and evening for seven days.)					
Skin Rash	**Ghole**	*Adenia cissampeloi-des*	Shrub	Wood	Lowland Forest
	Myonga pembe	?	Tree	Wood	Lowland Forest
	Mkula	*Pterocarpus sp.*	Tree	Bark	Forest
(Small piece of wood from **Ghole** amd **Myonga pembe** and bark of **Mkula** soaked in water overnight. Patient bathes in it morning and night for one month.)					
Snake Bite	***Kikulagembe***	?	Tree	Leaves	Lowland Forest
(Dry leaves burned and applied to wound immediately).					
Stomach Ache & Diarrhoea	***Mzumbasa***	*Ocimum suave*	Shrub	Roots	Shamba
	Mtura	*Solanum incanum*	Shrub	Roots	Shamba
(Roots of each boiled and water drunk. For diarrhoea, twenty leaves of ***Mvumbasa***, dried, powdered and drunk with water.)					
Stomach Ache	**Mshinga**	*Trema orientalis*	Tree	Roots	Shamba
	Mkwamab	*Securinega virosa*	Tree	Roots	Shamba
	Mtongwe	*Annona senegalensis*	Tree	Roots	Shamba
	Mkiika	?	Tree	Roots	Shamba
	Mbaazi	?	Shrub	Beans	Shamba
	Msasa	*Ficus exasperata*	Tree	Roots	Shamba
(Four roots of each boiled and water drunk.)					
Tuberculo-sis	***Mkwanga***	*Zanha golungensis*	Tree	Bark	Forest
(Burnt bark and ash snorted and eaten morning and evening for seven days.)					

Appendix 2: Medicinal Remedies used by Traditional Herbalists in Forest-Local Communities around Manga CGFR

Illness/ Problem	Vernacular Name	Botanical Name	Type	Part	Source
Asthma	**Mhasha**	*Veronica iodocalyx*	Bush	Leaves	Shamba
	Ushunguyuyu	*?*	?	Leaves	Shamba
(Squeeze sap from leaves and mix with coconut alcohol and drink.)					
Cataracts	**Ufyambo-fyambo**	*?*	?	Seeds	Shamba
(Eat one or two seeds three times a day for one week).					
Coughing	**Msasa**	*Ficus exasperata*	Tree	Leaves	Forest
(Grind leaves and dry. Add to porridge or tea once per day for two days.)					
Coughing	**Ufyambo-fyambo**	*?*	?	Leaves	Shamba
(Eat five to ten leaves three times a day for two days.)					
Constipation	**Ndeengwa**	*?*	?	Roots	Forest
	Mtamba kuzima	*Deinbolua borbonica*	Tree	Roots	Forest
(Roots boiled in water and drunk three times per day for three days.)					
Constipation and Gas	**Mtura**	*Solanum incanum*	Shrub	Roots	Forest
(Chew roots and swallow mucus.)					
Diarrhoea & Stomach Ache	**Mwati**	*Greenwayo-dndron suaveolens ssp. Usambari-cum*	Tree	Roots	Forest
	Mtonkwe	*?*	?	Roots	Forest
	Ingoingo	*?*	?	Roots	Forest
(Boil roots and drink water three times per day for three days.)					
Dysentery	**Sineakaya**	*?*	?	Leaves	Forest
	Ndugusi	*?*	?	Leaves	Forest
	Uvuvundi	*?*	?	Leaves	Forest
(Leaves ground and mixed with water and drunk three times a day for three days.)					
Fever in children	**Muuka**	*Microglossa densiflora*	Shrub	Roots & Leaves	Forest
	Mshwee	*?*	?	Roots & Leaves	Forest

	Hozandogoi	*Hyptis pectinata*	Shrub	Roots & Leaves	Shamba
(Roots boiled and water drunk three times a day for one day. Leaves of **Muuka** pounded and added with a little water and drunk and leaves of **Mshwee** and **Hozandogoi** squeezed for sap and drunk with a little water.)					
Gonorrhoea	**Kitupa mzitu**	?	?	Roots	Forest
(Chew roots and swallow mucus.)					
Indigestion	**Ukoka**	?	?	Leaves	Forest
(Boil leaves and drink.)					
Inducing Labour	**Msofu**	*Uvarioden-dron sp.*	Tree	Roots	Forest
(Boil roots and drink water. Must only be done just before the full moon.)					
Malaria	***Mchunga***	*Sonchus luxurians*	Herb	Leaves	Shamba
(Eat the leaves and drink the sap with a small amount of water. Only take once, when fever is present.)					
Menstrual Pain	**Mtindi**	?	Tree	Bark	Forest
(Boil bark and drink water three times a day for three days.)					
Menstrual Pain	**Utambaa mgoshwe**	?	?	Leaves	Forest
(Grind leaves, mix with water and drink.)					
Menstrual Pain	**Mtalawanda**	*Markhamia hildebrandtii*	Tree	Leaves	Forest
(Boil leaves and drink water.)					
Nausea	**Mwengee**	?	?	Leaves	Forest
	Mkwambe	?	Tree	Leaves	Forest
(Squeeze sap from leaves and drink with a little water.)					
Scabies	**Goe**	?	Climber	Roots	Forest
(Boil roots and bathe in water).					
Stomach Ache	**Mzumba-mbuku**	?	?	Leaves & Roots	Shamba
(Boil leaves and roots and drink water.)					

Bibliography

Aluma, J., Kahembwe, F. and Dubois, O. (1996), *Report on the Ugandan Round Table on Capacity Development for Sustainable Forestry in Africa: Understanding and Building on Capacities to Collaborate in Forestry*, FORI/NARO/IIED, October 1996 (unpublished).

Appleyard, B. (1992), *Understanding the present: Science and the Soul of Modern Man*, Pan Books, Picador, London.

Asia Forest Network (1995), *Transitions in Forest Management: Shifting Community Forestry from Project to Process*, Research Network Report, Centre for Southeast Asia Studies, Berkeley.

Asia Sustainable Forest Network (1994), *Policy Dialogue on Natural Forest Regeneration and Community Management*, Research Network Report, East West Centre, Hawaii.

Baland, J. M. and Platteau, J. P. (1996), *Halting Degradation of Natural Resources: Is There a Role for Rural Communities?* FAO and Clarendon Press, Oxford.

Baumann, O. (1889), 'Usambara', *Petermanns Geogr. Mitt.*, Vol. 35(2), pp. 41-47.

Bellville, A. (1875), 'Journey to the Universities' Mission Station of Magila on the Borders of the Usambara Country', *Proceedings of the Royal Geographical Society*, Vol. 20, pp. 74-78.

Bildsten, C. (1998), Personal Communication.

Borrini-Feyerabend, G. (1996), *Collaborative Management of Protected Areas: Tailoring the Approach to the Context*, Issues in Social Policy, IUCN, Gland.

Bromley, D. W. and Cernea, M. (1989), *The Management of Common Property Natural Resources: Some Conceptual Fallacies*, World Bank Discussion Paper 57, World Bank, Washington, D.C.

Bruce, J. W. (1993), 'Do Indigenous Tenure Systems Constrain Agricultural Development?', in T. J. Bassett and D. E. Crummey (eds), *Land in African Systems*, Wisconsin Press, Wisconsin.

Bruen, M. (1989), 'Hydrological Considerations of Development in the East Usambara Mountains', in A. C. Hamilton and R. Bensted-Smith (eds), *Forest Conservation in the East Usambara Mountains Tanzania*, IUCN, Gland.

Buchwald, J. (1897), 'Westusambara, die Vegetation und der Wirtschaftliche Werth der Landes', *Der Tropenflanzer*, Vol. 1(3), pp. 58-60, Vol. 1(4), pp. 82-85 and Vol. 1(5), pp. 105-108.

Burnett, G. W. and Kang'ethe, K. (1994), 'Wilderness and the Bantu Mind', *Environmental Ethics*, Vol. 16, pp. 145-160.

Cambridge-Tanzania Rainforest Project (1994), *A Biological and Human Impact Survey of the Lowland Forest, East Usambara Mountains, Tanzania*, Birdlife Study Report Number 59, Birdlife International, Cambridge.

Chambers, R. (1994a), 'The Origins and Practice of Participatory Rural Appraisal', *World Development*, Vol. 22(7), pp. 953-969.

Chambers, R. (1994b), 'Participatory Rural Appraisal (PRA): Challenges, Potentials and Paradigm', *World Development*, Vol. 22(10), pp. 1437-1454.

Chambers, R., Pacey, A. and Thrupp, L. A. (eds) (1989), *Farmer First: Farmer Innovation and Agricultural Research*, Intermediate Technology Publication, London.

Chauveau, J. P. (1996), 'La Logique des Systemes Coutumiers', in Lavigne Delville (ed.), *Foncier Rural,Ressources Renouvelables et Developpement: Analyse Comparative des Differentes Approches, GRET*, Novembre 1996, pp. 43-50.

Cherrett, I., O'Keefe, P., Heidenreich, A. and Middlebrook, P. (1995), 'Redefining the Roles of Environmental NGOs in Africa', *Development in Practice*, Vol. 5(1), pp. 26-35.

Cliffe, L., Luttrell, W. L. and Moore, J. E. (1975), 'The Development Crisis in Western Usambaras', in L. Cliffe, P. Lawrence, W. L. Luttrell, S. Migot-Adholla and J. S. Saul (eds) *Rural Co-operation in Tanzania*, Dar es Salaam.

Conte, C. (1996), 'Nature Reorganised: Ecological History in the Plateau Forests of the West Usambara Mountains, 1850-1935', in G. Maddox, J. Giblin, and I. N. Kimambo (eds). *Custodians of the Land: Ecology and Culture in the History of Tanzania*, James Currey, London.

Cory, H. (1962a), 'The Sambaa Initiation Rites for Boys', *Tanganyika Notes and Records*, Vol. 58, pp. 2-7.

Cory, H. (1962b), 'Tambiko (Fika)', *Tanganyika Notes and Records,* Vol. 59, pp. 274-82.

Cunneyworth, P. (1996a), *Kambai Forest Reserve: A Biodiversity Survey*, East Usambara Catchment Forest Project Technical Paper 35, Forestry and Beekeeping Division and Finnish Forest and Park Service and Society for Environmental Exploration, Dar es Salaam, Vantaa and London.

Cunneyworth, P. (1996b), *Semdoe Forest Reserve: A Biodiversity Survey*, East Usambara Catchment Forest Project Technical Paper 36, Forestry and Beekeeping Division and Finnish Forest and Park Service and Society for Environmental Exploration, Dar es Salaam, Vantaa and London.

Cunneyworth, P. and Stubblefield, L. (1996), *Magoroto Forest: A Biodiversity Survey*, East Usambara Catchment Forest Project Technical Paper 30, Forestry and Beekeeping Division and Finnish Forest and Park Service and Society for Environmental Exploration, Dar es Salaam, Vantaa and London.

de Beer and McDermott (1996), *The Economic Value of Non-Timber Forest Products in Southeast Asia*, NC-IUCN, Amsterdam.

Dobson, E. B. (1940), 'Land Tenure of the Wasambaa', *Tanganyika Notes and Records*, Vol. 10, pp. 1-27.

Doring, P. (1899), *Morgendammerung in Deutsche-Ostafrika*, Ein Rundgang durch die Ostafrikanische Mission, Berlin.

Dubois, O. (1997), *Rights and Wrongs of Rights to Land and Forest Resources in Sub-Saharan Africa: Bridging the Gap between Customary and Formal Rules*, Forestry and Land Use Programme, IIED, London.

East Usambara Catchment Forest Project (1995), *Project Document, Phase 2: 1995-98, Volumes 1 and 2*, Ministry of Tourism, Natural Resources and Environment, Dar es Salaam.

The Economist (1995), 'A Matter of Title', 9 December.

Egger, K. and Glaeser, B. (1975), *Politische Ökologie der Usambara-Berge in Tanzania*, Nairobi.

Eick, E. (1896), 'Bericht über meine Reise ins Kwai und Masumbailand (Usambara) vom 12 bis 16.März 1896', *Mitt. Deutschen Schutzgeb.*, Vol. 9(3), pp. 184-188.

Ekins, P. (1992), *Wealth Beyond Measure: An Atlas of New Economics*, Gaia Books, London.

Emerton, L. (1996), *Local Livelihoods and Forest Biodiversity Loss: A Case from Kenya*, paper prepared for the workshop on the Economics of Biodiversity Loss, IUCN, Gland.

Evers, Y. D. (1994), *Subsistence Strategies and Wild Resource Utilisation: Pugu Forest Reserve, Tanzania*, based on MSc Thesis, Department of Anthropology, University College London, London (unpublished).

Fairhead, J. (1992), *Indigenous Technical Knowledge and Natural Resources Management in Sub-Saharan Africa: A Critical Review*, paper prepared for the Social Science Research Council Project on African Agriculture, Dakar, January.

Fairhead, J. and Leach, M. (1996a), *Misreading the African Landscape: Society and Ecology in a Forest-Savanna Mosaic*, Cambridge University Press, Cambridge.

Fairhead, J. and Leach, M. (1996b), 'Rethinking the Forest-Savanna Mosaic: Colonial Science and its Relics in West Africa', in M. Leach and R. Mearns (eds), *The Lie of the Land: Challenging Received Wisdom on the African Environment*, Villiers Publications, London.

Farler, J. P. (1879), 'The Usambara Country in East Africa', *Proceedings of the Royal Geographical Society*, Vol. 1(2), pp. 81-97.

Feierman, S. (1974), *The Shambaa Kingdom: A History*, University of Wisconsin Press, Madison, WI.

Feierman, S. (1990), *Peasant Intellectuals: Anthropology and History in Tanzania*, University of Wisconsin Press, Madison, WI.

Fisher, R. (1995), *Collaborative Management of Forests for Conservation and Development*, issues in Forest Conservation, IUCN, Gland.

Fleuret, A. (1979a), 'The Role of Wild Foliage Plants in the Diet: A Case Study from Lushoto, Tanzania', *Ecology of Food and Nutrition*, Vol. 8, pp. 87-93.

Fleuret, A. (1979b), 'Methods of Evaluation of the Role of Fruits and Wild Greens in Shambaa Diet: A Case Study', *Medical Anthropology*, Vol. 3, pp. 249-269.

Fleuret, A. (1980), 'Non-food Uses of Plants in Usambara', *Economic Botany*, Vol. 34(4), pp. 320-333.

Fleuret, P. (1978), *Farm and Market: A Study of Society and Agriculture in Northern Tanzania*, Unpublished PhD thesis, University of California.

Forest Department of Tanganyika Territory (1930), *Annual Report*, Dar es Salaam.

Förster, B. (1890), *Deutsch-Ostafrika. Geographie und Geschichte der Colonie. Mit einer Karte von Deutsch-Ostafrika*, Leipzig.

Fortmann, L. (1985), 'The Tree Tenure Factor in Agroforestry, with Particular Reference to Africa', *Agroforestry Systems*, Vol. 2, pp. 229-251.

Franks, P. (1995), *What Role for Communities in PA Management in Uganda?* CARE, Uganda.

Funtowicz, S. O. and Ravetz, J. R. (1990), *Global Environmental Issues and the Emergence of Second Order Science*, Commission of the European Communities, Luxenburg.

Government of Tanganyika (1945), *The Forest Policy of Tanganyika*, Forest Department Annual Report, Dar es Salaam.

Government of the United Republic of Tanzania (1989), *Tanzania Forestry Action Plan 1990-2000*, Ministry of Lands, Natural Resources and Tourism, Dar es Salaam.

Grant, D. K. S. (1924), 'Forestry in Tanganyika,' *Empire Forestry*, Vol. 3, pp. 33-38.

Griffiths, C. J. (1993), 'The Geological Evolution of East Africa', in J. C. Lovett and S. K. Wasser (eds), *Biogeography and Ecology of the Rain Forests of Eastern Africa*, Cambridge University Press, Cambridge.

Hamilton, A. C. (1988), 'Conservation of the East Usambara Forests, with emphasis on Biological Conservation', *Acta Univ. Ups. Symb. Bot. Ups.*, Vol. 28(3), pp. 244-254.

Hamilton, A. C. and Bernsted-Smith, R. (eds) (1989), *Forest Conservation in the East Usambara Mountains, Tanzania*, IUCN, Gland.

Härkönen, M., Saarimaki, T. and Mwasumbi, L. (1995), 'Edible Mushrooms of Tanzania, *Karstenia*, Vol. 35.

Harmon, D. (1991), 'National Park Residency in Developed Countries: The Example of Great Britain', in P. C. West and S. R. Brechin (eds), *Resident Peoples and National Parks*, University of Arizona Press, Tucson, Arizona.

Hermansen, J. E., Benedict, F., Corneliussen, T., Hofsten, J. and Venvik, H. (1985), *Catchment Forestry in Tanzania: Status and Management*, Consultancy Report for NORAD, Oslo.

Hesseling, G. and Ba, B. M. (1994), *Land Tenure and Natural Resource Management in the Sahel: Experiences, Constraints and Prospects*, Regional Synthesis, Regional Conference on Land Tenure and Decentralisation in the Sahel. CILSS (Club de Sahel), Praia, Cape Verde.

Hisham, M. A., Sharma, J., Ngaiza, A., and Atampugre, N. (1991), *Whose Trees? A People's View of Forestry Aid*, Panos Publications Ltd., London.

Hobley, M. (1995), *Institutional Change within the Forest Sector: Centralised Decentralisation*, Rural Development Forestry Network, ODI, January 1995.

Hobley, M. (1996), 'Participatory Forestry: The Process of Change in India and Nepal', *Rural Development Study Guide* 3, Overseas Development Institute and Rural Development Forestry Network, London.

Hobley, M. and Shah, K. (1996), 'What Makes a Local Organisation Robust? Evidence from India and Nepal', *Natural Resource Perspectives*, 11(July), Overseas Development Institute, London.

Holst, C. (1893), 'Der Landbau der Eingeborenen von Usambara', *Deutsche Kolonialzeitung*,Vol. 6(9), pp. 113-114 and Vol. 6(10), pp. 128-130.

Howell, K. M. (1989), 'The East Usambara Fauna', in A. C. Hamilton and R. Bensted-Smith (eds), *Forest Conservation in the East Usambara Mountains Tanzania*, IUCN, Gland.

IUCN (1997), *Non-Timber Forest Products from the Tropical Forests of Africa: A Bibliography*, NC-IUCN, Amsterdam.

Iversen, S. T. (1991), *The Usambara Mountains, NE Tanzania: History, Vegetation and Conservation*, Uppsala Universitet, Uppsala.

Joelson, F. S. (1928), 'Tanganyika Territory, Eastern Africa Today, East Africa (A London Newspaper)', in Tanzania Forest Conservation Group (1997), *The Arc Journal*, Vol. 6, Tanzania Forest Conservation Group, Dar es Salaam.

Johansson, L. (1991), *Successful Tree Growers: Why People Grow Trees in Babati District, Tanzania*, Working paper 155, International Rural Development Centre, Uppsala.

Johansson, S. G. (1994), *Forest Conservation in the East Usambara Mountains: A Map Supplement*, East Usambara Catchment Forest Project, Working Paper 1, Forest and Beekeeping Division and Finnish Forest and Park Service, Dar es Salaam and Vantaa.

Johansson, S. G. and Sandy, R. (1996), *Protected Areas and Public Lands: Land Use in the East Usambara Mountains*, East Usambara Catchment Forest Project Technical Paper 28, Forestry and Beekeeping Division & Finnish Forest and Park Service, Dar es Salaam & Vantaa.

Johnston, K. (1879), 'Notes of a Trip from Zanzibar to Usambara, in February and March 1879, *Proceedings of the Royal Geographical Society*, Vol. 1(9), pp. 545-558.

Kajembe, G. C. (1994), *Indigenous Management Systems as a Basis for Community Forestry in Tanzania: A Case Study of Dodoma Urban and Lushoto Districts*, Wageningen Agricultural University, Wageningen.

Kajembe, G. C. and Mwaseba, D. (1994), *The Extension and Communication Programme for the East Usambara Catchment Forest Project*, East Usambara Catchment Forest Project, Technical Paper No. 11, Forestry and Beekeeping Division, Finnish Forest and Park Service, Dar es Salaam and Vantaa.

Kalaghe, A. G., Msangi, T. H. and Johansson, L. (1988), 'Conservation of Catchment Forests in the Usambara Mountains', *J. Tanz. Ass. For.*, Vol. 6, pp. 37-47, Dar es Salaam.

Kessey, J. F. and O'Kting'ati, A. (1994), 'An Analysis of some Socio-economic Factors Affecting Farmers' Involvement in Agroforestry Extension Projects in Tanzania', *Annals of Forestry*, Vol. 2 (1), pp. 26 -32.

Kimambo, I. N. (1996), 'Environmental Control and Hunger: In the Mountains and Plains of Northeastern Tanzania', in G. Maddox, J. Giblin, and I. N. Kimambo, (eds), *Custodians of the Land: Ecology and Culture in the History of Tanzania*, James Currey, London.

Korten, D. C. (1984), 'People Centered Development: Towards a Framework', in D. C. Korten and R. Klauss (eds), *People-Centered Development*, Kumarian Press, West Hartford, Connecticut.

Krapf, J. L. (1858), *Reisen in Ostafrika Ausgeführt in den Jahren 1837-1855: Meine Grössere Reisen in Ostafrika*, Stuttgart.

Lagerstedt, E. (1994), *Views, Needs, Uses and Problems Connected with the Forest in Kazimzumbwi Village, Tanzania*, Coastal Forest Project, Wildlife Conservation Society of Tanzania, Dar es Salaam.

Lamphear, J. (1970), 'The Kamba and the Northern Mrima Coast', in R. Gray and D. Birmingham (eds), *Precolonial African Trade*, Oxford University Press, London, New York and Nairobi.

Leach, M. (1991), 'Engendered Environments: Understanding Natural Resource Management in the West African Forest Zone', *IDS Bulletin*, Vol. 22(4), pp. 17-24.

Leach, M. and Mearns, R. (eds) (1996), *The Lie of the Land: Challenging Received Wisdom on the African Environment*, Villiers Publications, London.

Leach, M., Mearns, R. and Scoones, I. (1997), 'Environmental Entitlements: A Framework for Understanding the Institutional Dynamics of Environmental Change, *IDS Discussion Paper*, 359, Institute of Development Studies, Brighton.

Leach, M., Mearns, R. and Scoones, I. (1999), 'Environmental Entitlement: Dynamics and Institutions in Community-Based Natural Resource Management, *World Development*, Vol. 27(2), pp. 225-247.

Legislative Council of Tanganyika (1953), *Forest Policy*, Sessional paper No. 1, pp. 1-5, Dar es Salaam.

Lindstrom, J. and Kingamkono, R. (1991), *Foods from Forest, Fields and Fallows: Nutritional and Food Security Roles of Gathered Food and Livestock Keeping in Two Villages in Babati District, Northern Tanzania*, Working Paper No. 184, Swedish University of Agricultural Sciences, International Rural Development Centre, Uppsala.

Litterick, M. (1989), 'Assessment of Water Quality of the Sigi River', in A. C. Hamilton and R. Bensted-Smith (eds), *Forest Conservation in the East Usambara Mountains Tanzania*, IUCN, Gland.

Locke, C. (1999), 'Constructing a Gender Policy for Joint Forest Management in India', *Development and Change*, Vol. 30, pp. 265-285.

Lovett, J. C. and Congdon, T. C. E. (1990), 'Notes on Lulanda Forest, Southern Udzungwa Mountains', *East Africa Natural History Society Bulletin*, Vol. 20, pp. 21.

Lovett, J. C. and Pocs, T. (1992), *Catchment Forest Reserves of Iringa Region: A Botanical Appraisal*, Dar es Salaam (unpublished).

Lovett, J. C. and Wasser, S. K. (eds) (1993), *Biogeography and Ecology of the Rain Forests of Eastern Africa*, Cambridge University Press, Cambridge.

Lundgren, B. (1978), 'Soil Conditions and Nutrient Cycling Under Natural and Plantation Forests in Tanzanian Highlands', *Rep. For. Ecol. For. Soils*, Vol. 31, pp. 1-426.

Lundgren, L. (1985), *Catchment Forestry in Tanzania*, A report prepared for the joint Tanzanian/Swedish review 1984, revised for the joint Tanzanian/Nordic Forestry Sector review 1985, Dar es Salaam (unpublished).

Lynch, O. (1992), *Securing Community-Based Tenurial Rights in the Tropical Forests of Asia: An Overview of Current and Prospective Strategies*, WRI Issues in Development, November 1992.

McKean, M. and Ostrom, E. (1995), 'Common Property Regimes in the Forest: Just a Relic from the Past?', *Unasylva*, Vol. 46(180), pp. 3-15.

Madumulla, J. S. (1995), *Proverbs and Sayings: Theory and Practice*, Institute of Kiswahili Research, University of Dar es Salaam, Dar es Salaam.

Matthews, P. (1994), *Medicinal Plants of the Tanzanian Coastal Forests: A List of Species with Local Names and Applications*, The Society for Environmental Exploration and University of Dar es Salaam, London and Dar es Salaam.

Mbiti, J. S. (1970), *African Religions and Philosophy*, Anchor Books, Garden City, New York.

Mearns, R. (1991), *Structural Adjustment and the Environment: Reflections on Scientific Method*, IDS Discussion Paper 284, Institute of Development Studies, University of Sussex, Brighton.

Meshack, C. and Woodcock, K. A. (1998), *Simuliza Kutoka Kijiji cha Lulanda: Watu, Misitu na Miti*, Tanzania Forest Conservation Group, Dar es Salaam.

Messerschmidt, D. A. (ed), (1993), *Common Forest Resource Management: Annotated Bibliography of Asia, Africa and Latin America*, FAO Community Forestry Note 11, FAO, Rome.

Meyer, H. (1914), 'Das Deutsche Kolonialreich', *Ostafrika*, Vol. 1(1).

Meyer, H. and Baumann, O. (1888), 'Dr. Hans Meyer's Usambara-Expedition', *Mitt. Deutschen Schutzgeb.*, Vol. 1, pp. 199-205.

Milne, G. (1937), 'Essays in Applied Pedology: Soil Type and Soil Management in Relation to Plantation Agriculture in the East Usambara', *E. Afr. Agr. J.*, Vol. 3(1), pp. 7-20.

Mogaka, H. R. (1991), *Local Utilisation of Arabuko-Sokoke Forest Reserve,* Interim Socio-economic Report 6, Indigenous Forest Conservation Project, Forest Department, Nairobi.

Moreau, R. E. (1935), 'A Synecological Study of Usambara, Tanganyika Territory, with Special Reference to Birds', *J. Ecol.*, Vol. 23, pp. 1-43.

Mortimore, M. (1996), 'Land Tenure and Resource Access in West Africa', in Lavigne Delville (ed), *Foncier Rural Ressources Renouvelables et Developpement, GRET,* Novembre 1996.

Muir, C. (1998), Personal Communication.

Nhira, C. and Matose, F. (1996), *Joint Forest Management and Resource Sharing: Lessons from India and Zimbabwe*, Forest Participation Series 5, IIED, London.

Nipashe, (1996), February, *Wagombea Ardhi ya Kilomita Sita*, Dar es Salaam.

Norton-Griffiths, M. and Southey, C. (1995), 'The Opportunity Costs of Biodiversity in Kenya', *Ecological Economics*, Vol. 12, pp. 125-139.

Owen, M. (1992), *East Usambara Conservation and Development Project: Forest Products Survey*, Bellerive Foundation, Nairobi.

Parry, M. S. (1962), 'Progress in the Protection of Stream-Source Areas in Tanganyika', in H. C. Pereira (ed.), 'Hydrological Effects of Changes in Land Use in Some East African Catchment Areas', *E. Afr. Agr. For. J.* (Special issue), Vol. 27, pp. 104-106.

Pattnaik, B. K. and Dutta, S. (1997), 'JFM in South-West Bengal: A Study in Participatory Development', *Economic and Political Weekly*, December 13, pp. 3225-3232.

Peters, T. (1987), *Thriving on Chaos: Handbook for a Management Revolution*, Alfred A. Knopf, New York.

Pimbert, M. P. and Pretty, J. N. (1997), 'Parks, People and Professionals: Putting "Participation" into Protected Area Management', in K. B. Ghimire and M. P. Pimbert (eds), *Social Change and Conservation: Environmental Politics and Impacts of National Parks and Protected Areas*, Earthscan Publications Limited, London.

Platteau, J. P. (1996), 'The Evolutionary Theory of Land-Rights as Applied to Sub-Saharan Africa: A Critical Assessment', *Development and Change*, Vol. 27, pp. 29-86.

Poffenberger, M. and McGean, B. (eds) (1996), *Village Voices, Forest Choices: Joint Forest Management in India*, Oxford University Press, Delhi.

Poffenberger, M. with A. Bhattachacharya, A. Khare, A. Rai, S. Roy, N. Singh, and K. Singh (1996), *Grassroots Forest Protection: Eastern Indian Experiences*, no. 7, March, Asia Forest Network, Centre for Southeast Asia Studies, Berkeley, California.

Pretty, J. N. (1993), *Participatory Inquiry and Agricultural Research*, IIED, London.

Pretty, J. N. (1994), 'Alternative Systems of Inquiry for Sustainable Agriculture', *IDS Bulletin*, Vol. 25(2), pp. 37-48.

Pretty, J. N., Guijt, I., Thompson, J. and Scoones, I. (1995), *Participatory Learning and Action: A Trainers Guide*, Participatory Methodology Series, IIED, London.

Raharimalala, S. R. (1996), *Les Conventions Coutumieres Obligatoires des Collectivites Villageoises (DINA) dans la Gestion Communautaire des Ressources Naturelles a Madagascar*, Communication presentee lors du Colooque Panafricain sur la Gestion Communautaire des Ressources Naturelles et le Developpement Durable, Harare, Zimbabwe.

Ranzanaka, S. J. (1996), *La Place du 'Dina' dans la Strategie de Gestion Communautaire des Ressources Renouvelables et le Suivi des Feux de Vegetation dans le Sud-Ouest de Madagascar*, Communication presentee lors du Colloque Panafricain sur la Gestion Communautaire des Ressources Renouvelables et le Developpement Durable, Harare, Zimbabwe.

Redmayne, A. (1968), 'The Hehe', in A. Roberts (ed.), *Tanzania Before 1900*, East African Publishing House, Dar es Salaam.

Ribot, J. C. (1995), *Local Forest Control in Burkino Faso, Mali, Niger, Senegal and the Gambia: A Review and Critique of New Participatory Policies*, Regional synthesis report of the World Bank, World Bank, Washington, D.C.

Richards, P. (1985), *Indigenous Agricultural Revolution: Ecology and Food Production in West Africa*, Allen and Unwin, Hemel Hempstead.

Rodgers, W. A. (1993), 'The Conservation of the Forest Resources of East Africa: Past Influences, Present Practices and Future Needs', in J. C. Lovett, and S. K. Wasser (eds), *Biogeography and Ecology of the Rainforests of Eastern Africa*, Cambridge University Press, Cambridge.

Rodgers, W. A. (1996), *Patterns of Loss of Forest Biodiversity: A Global Perspective*, presented at the workshop on the economics of biodiversity loss, IUCN, Gland.

Rodgers, W. A. (1997), *Paper Presented at Eastern Arc Conference*, Morogoro (unpublished).

Rodgers, W. A. and Homewood, K. M. (1982), 'Biological Values and Conservation Prospects for the Forests and Primate Populations of the Udzungwa Mountains, Tanzania', *Biological Conservation*, Vol. 24, pp. 285-304.

Routledge, W. S. and Routledge, K. (1910), *With a Prehistoric People: The Akikuyu of British East Africa*, Edward Arnold, London.

Ruffo, C. K. (1989), 'Some Useful Plants of the East Usambaras', in A. C. Hamilton and R. Bernsted-Smith (eds), *Forest Conservation in the East Usambara Mountains, Tanzania*, IUCN, Gland.

Ruffo, C. K., Mwasha, I. V. and Mmari, C. (1989), 'The Use of Medicinal Plants in the East Usambaras', in A. C. Hamilton and R. Bernsted-Smith (eds), *Forest Conservation in the East Usambara Mountains, Tanzania*, IUCN, Gland.

Sayer, J. A., Harcourt, C. S. and Collins, N. M. (1992), *The Conservation Atlas of Tropical Forests: Africa*, Macmillan, London.

Schabel, H. (1990), 'Tanganyika Forestry Under German Colonial Administration, 1891-1991', *Forest and Conservation History*, Vol. 34(3), p. 131.

Scheffler, G. (1901), 'Über die Beschaffenheit des Usambara-Urwaldes und über den Laubwechsel an Bäumen Desselben', *Not. Kön. Bot. Gart. Mus. Berlin*, Vol. 3(27), pp. 139-166.

Scoones, I., Melnyk, M. and Pretty, J. N. (1992), *The Hidden Harvest: Wild Foods and Agricultural Systems, A Literature Review and Annotated Bibliography*, IIED/SIDA/WWF, London.

Senge, P. M. (1992), *The Fifth Discipline: The Art and Practice of the Learning Organisation*, Century Business, London.

Sharma, N., Reitbergen, S., Heimo, C. and Patel, J. (1994), *A Strategy for the Forest Sector in Sub-Saharan Africa*, World Bank Technical Paper 251, Africa Technical Department Series, World Bank, Washington, D.C.

Shepherd, G. (1991), 'The Communal Management of Forests in the Semi-Arid and Sub-Humid Regions of Africa: Past Practice and Prospects for the Future', *Development Policy Review*, Vol. 9, pp. 151-176.

Shepherd, G., Kiff, L. and Robertson, D. (1995), *The Importance of Common Property Issues, Tenure and Access Rights in Relation to Land Use Management and Planning at the Forest-Agriculture Interface*, Literature Review for the Natural Resources Systems Programme, NRI/ODA, London (unpublished).

Siebenlist, T. (1914), *Forstwirtschaft in Deutsch-Ostafrika*, Berlin.

Talbott, K. and Khadka, S. (1994), *Handing It Over: An Analysis of the Legal and Policy Framework of Community Forestry in Nepal*, Cases in Development, World Resources Institute, New York.

Tanzania Forest Conservation Group (1983-1998), *General Files*, Dar es Salaam (unpublished).

Tanzania Forest Conservation Group (1993-1998), *Lulanda Field Report Files*, Dar es Salaam (unpublished).

Taylor, M. (1997), 'Governing Natural Resources', *Society and Natural Resources*, Vol. 11, pp. 251-258.

Troup, R. S. (1936), *Report on Forestry in Tanganyika Territory*, Forest Department, Dar es Salaam.

Tye, A. (1993), *Magoroto Rainforest Conservation: Proposal for Establishment of New Reserve*, Amani (unpublished).

The United Republic of Tanzania (1994), *Report of the Presidential Commission of Inquiry into Land Matters, Vol. 1: Land Policy and Land Tenure Structure*, The Ministry of Lands, Housing and Urban Development, Government of the United Republic of Tanzania, in co-operation with, the Scandinavian Institute of African Studies, Uppsala.

The United Republic of Tanzania (1998), *National Forest Policy*, Ministry of Natural Resources and Tourism, Dar es Salaam.

Vainio-Mattila, K., Mwasumbi, L. and Lahti, K. (1997), *Traditional Use of Wild Vegetables in the East Usambara Mountains*, East Usambara Catchment Forest Project Technical Report 37, Forestry and Beekeeping Division and Finnish Parks Service, Dar es Salaam and Vantaa.

Vest, J. H. C. (1991), 'The Concept of Wilderness: A Propriety Right Over the Land', *Western Wildlands*, Vol. 17(2), pp. 2-6.

Volkens, G. (1897), 'Zur Frage der Aufforstung in Deutsch-Ostafrika', *Notizbl. Kön. Bot. Gart. Mus. Berlin*, Vol. 2 (11), pp. 12-20.

Warburg, O. (1894), 'Die Kulturpflanzen Usambaras', *Mitt. Deutschen Schutzgeb.*, Vol. 7, pp. 131-199.

Watson, J. R. (1972), 'Conservation Problems, Policies and the Origins of the Mlalo Basin Rehabilitation Scheme, Usambara Mountains, Tanzania', *Geogr. Ann., ser. A.*, Vol. 54(3/4), pp. 221-226.

Wells, M. and Brandon, K. with Hannah, L. (1993), *People and Parks: Linking Protected Area Management with Local Communities*, The International Bank for Reconstruction and Development, World Bank, Washington, D.C.

West, P. C. and Brechin, S. R. (eds) (1991), *Resident Peoples and National Parks*, University of Arizona Press, Tucson, Arizona.

Wily, L. (1995a), *Village Forest Reserves in the Making: The Story of Duru-Haitembo*, International Rural Development Centre, Uppsala.

Wily, L. (1995b), *Guidelines for Facilitation: Helping Villagers Manage Their Own Forests*, ORGUT, Dar es Salaam.

Wily, L. (1996), *Villagers as Forest Managers: The Case of Duru-Haitemba and Mgori Forest in Tanzania*, Forests, Trees and People Programme, Working Paper Series, FAO, Rome.

Wily, L. (1997), *Finding the Right Institutional and Legal Framework for Community-Based Natural Forest Management: The Tanzanian Case*, Centre for International Forestry Research, Jakarta.

Wily, L. (1999), 'Moving Forward in African Community Forestry: Trading Power, Not Use Rights', *Society and Natural Resources*, Vol. 12, pp. 49-61.

Winans, E. V. (1962), *Shambala: The Constitution of a Traditional State*, Berkeley.

Wohltmann, F. (1902), 'Die Aussichten des Kaffeebaues in den Usambara-Bergen', *Der Tropenpflanzer*, Vol. 6(12), pp. 612-616.

Woodcock, K. A. (1995), *Indigenous Knowledge and Forest Use: Two Case Studies from the East Usambaras, Tanzania*, East Usambara Catchment Forest Project Technical Report 22, Forestry and Beekeeping Division and Finnish Parks Service, Dar es Salaam and Vantaa.

World Bank (1992), *Empowering Villages to Manage their Natural Resources: Rural Land Policy in Tanzania*, Southern Africa Department, Agriculture Operations Division, World Bank, Washington, D.C.

Yadama, G. N. (1997), 'Tales from the Field: Observations on the Impact of Non-Governmental Organisations', *International Social Work*, Vol. 40, pp. 145-162.

Index

For Product Safety Concerns and Information please contact our EU representative GPSR@taylorandfrancis.com Taylor & Francis Verlag GmbH, Kaufingerstraße 24, 80331 München, Germany

T - #0119 - 160425 - C0 - 216/149/11 - PB - 9781138728479 - Gloss Lamination